—THREE VIEWS— ON CREATION AND EVOLUTION

Books in the Counterpoints Series

Are Miraculous Gifts for Today?
Five Views on Law and Gospel
Five Views on Sanctification
Four Views on Hell
Four Views on Salvation in a Pluralistic World
Four Views on the Book of Revelation
Three Views on Creation and Evolution
Three Views on the Millennium and Beyond
Three Views on the Rapture
Two Views on Women in Ministry

Stanley N. Gundry (S.T.D., Lutheran School of Theology at Chicago) is vice president and editor-in-chief of the Book Group at Zondervan. He graduated summa cum laude from both the Los Angeles Baptist College and Talbot Theological Seminary before receiving his Masters of Sacred Theology from Union College, University of British Columbia. With more than thirty-five years of teaching, pastoring, and publishing experience, he is the author or coauthor of numerous books and a contributor to numerous periodicals.

COUNTERPOINTS

THREE VIEWS ON CREATION AND EVOLUTION

J. P. Moreland
and
John Mark Reynolds
General Editors

GRAND RAPIDS, MICHIGAN 49530

Three Views on Creation and Evolution

Requests for information should be addressed to:

Zondervan, *Grand Rapids, Michigan 49530*

Library of Congress Cataloging-in-Publication Data

Three views on creation and evolution / J. P. Moreland and John Mark Reynolds, general editors ; contributors, Paul Nelson ... [et al.] ; respondents, John Jefferson Davis ... [et al.], First.
p. cm.—(Counterpoints)
Includes bibliographical references and index.
ISBN 0-310-22017-3
1. Creationism. 2. Evolution—Religious aspects—Christianity.
I. Moreland, James Porter, 1948–. II. Reynolds, John Mark, 1963–.
III. Nelson, Paul. IV. Davis, John Jefferson. V. Series: Counterpoints (Grand Rapids, Mich.)
BS651.T485 1999
231.7'652—dc21 98-33239

Printed in the United States of America

02 03 04 05 06 / DC / 10 9 8 7 6

CONTENTS

INTRODUCTION

J. P. Moreland and John Mark Reynolds

One of the most important influences shaping the modern world is science. People who lived during the Civil War had more in common with Abraham than with us. From DNA research and space travel to computer science and neurophysiology, ours is a world of science.

For Christians, this comes as no surprise. We believe in a rational God who created the world. We also believe that he created our human intellectual faculties in such a way that we could gain knowledge from our exploration of the world in which he has placed us. But for many people, regardless of whether they are right or wrong, the relationship between science and Christian theology is not this simple. In addition, many people widely disagree about the integration of science and theology, especially how Christians ought to view evolution. For some of them, science and Christian theology have little or nothing to do with each other. Others claim that a number of recent scientific discoveries have provided intellectual support for certain theological claims, including, for example, that the universe had a beginning; that it is fine-tuned; that the degree and type of complexity found in biological systems could not—or most likely did not—evolve from a mindless process in which matter rearranged according to natural law and chance; or that psychological research requires us to treat human beings as deeply unified selves with the power of free agency.

Another large group of people, however, embrace a third view of science and theology—in some way or another, science has falsified Christianity in general or various doctrinal commitments in particular, or at least science has shown that Christian teaching is, intellectually speaking, vastly inferior to science and therefore must be accepted by blind faith for those with certain dispositions. As philosophical naturalist Wilfred Sellars put it, "In the dimension of describing and explaining the world, science is the measure of all things, of what is that it is, and of what is not that it is not."[1] Nowhere is this attitude more prevalent than the topic of creation and evolution. As one atheist proclaimed, "Evolution has made the world safe for atheists."

Because Christians are interested in the truth and because they are called to proclaim and defend their views to an unbelieving world, it is important for the believing community to think carefully about how to integrate their carefully formed theological beliefs with a careful evaluation of the "deliverances" of science, especially in the area of creation and evolution. As St. Augustine noted long ago, "We must show our Scriptures not to be in conflict with whatever [our critics] can demonstrate about the nature of things from reliable sources."[2] Similarly, those outside the church have an intellectual duty to show that their views on various topics do not conflict with a rationally formed and justified theology when it is relevant to do so.

In light of this need for integration, the purpose of this book is to inform the reader about the issues involved in the dialogue about creation and evolution among three different schools of thought: (1) young earth or recent creation; (2) progressive or old earth creation; and (3) theistic evolution or evolving creation.[3] In the introduction to follow, we hope to orient you to the topics of importance in this dialogue by discussing different models for integrating science and theology, an important challenge to theology from what is called scientism, the main issues

---

[1]Wilfred Sellars, *Science, Perception, and Reality* (London: Routledge & Kegan Paul, 1963), 173.

[2]Augustine, *De genesi ad litteram* 1.21. Cited in Ernan McMullin, "How Should Cosmology Relate to Theology?" in *The Sciences and Theology in the Twentieth Century*, ed. Arthur R. Peacocke (Notre Dame, Ind.: University of Notre Dame Press, 1981), 20.

[3]See also Del Ratzsch, *The Battle of Beginnings* (Downers Grove, Ill.: InterVarsity Press, 1996).

involved in the three-way dialogue about creation and evolution, and a historical overview of the debate, as well as how we will proceed in the remainder of the book.

THE INTEGRATION OF THEOLOGY AND OTHER DISCIPLINES

Different models for integrating theology and other disciplines have been proposed, among which are the following five views:[4]

1. The Two Realms View

Propositions, theories, or methodologies in theology and another discipline may involve two distinct, nonoverlapping areas of investigation. For example, debates about angels or the extent of the atonement have little to do with organic chemistry. Similarly, it is of little interest to theology whether a methane molecule contains three or four hydrogen atoms.

2. The Complementarity View

Propositions, theories, or methodologies in science and theology are noninteracting, complementary approaches to the same reality that adopt very different standpoints, ask and answer very different kinds of questions (e.g., evolutionary science tells us what happened and how while theology tells us who guided the process and why), involve different levels of description, employ very different cognitive attitudes (e.g., objectivity and logical neutrality in science, personal involvement and commitment in theology), and/or are constituted by very different language games. These different, authentic perspectives are partial and incomplete and, therefore, must be integrated into a coherent whole. However, each level of description is complete at its own level without having gaps at that level for the other perspective to fill and without having the possibility of direct competition and conflict. There may be implications for

[4]See A. R. Peacocke, "Introduction," in *The Sciences and Theology in the Twentieth Century*, ed. Arthur R. Peacocke (Notre Dame, Ind.: University of Notre Dame Press, 1981), xiii–xv.

theology from science or vice versa, but these are rare and should allow theology and science to remain in their own proper domains with no illicit territory encroachment from one to the other. Some form of theistic evolution is the favored model of creation-evolution integration for complementarians.

3. The Direct Interaction View

Propositions, theories, or methodologies in theology and another discipline may directly interact in such a way that one area of study either offers rational support for the other or raises rational difficulties for the other. For example, certain theological teachings about the existence of the soul raise rational problems for philosophical or scientific claims that deny the existence of the soul. The general theory of evolution raises various difficulties for certain ways of understanding the book of Genesis. Some have argued that the Big Bang theory tends to support the theological proposition that the universe had a beginning.

4. The Presuppositional View

Theology tends to support the presuppositions of another discipline and vice versa. Some have argued that many of the presuppositions of science (e.g., the existence of truth, the rational orderly nature of reality, and the adequacy of our sensory and cognitive faculties as tools suited for knowing the external world) make sense and are easy to justify given Christian theism, but are odd and without ultimate justification in a naturalistic worldview. Similarly, some have argued that philosophical critiques of epistemological skepticism and defenses of the existence of a real theory-independent world and a correspondence theory of truth offer justification for some of the presuppositions of theology.

5. The Practical Application View

Theology fills out and adds details to general principles in another discipline and vice versa, and theology helps one practically apply principles in another discipline and vice versa. For example, theology teaches that fathers should not provoke their children to anger and psychology can add important details about what this means by offering information about family systems and the nature and causes of anger. Psychology can devise various tests for assessing whether one is a mature person and theology can offer a normative definition to psychology as to what a mature person is.

By way of application to the creation-evolution dialogue, Christians are divided about the proper approach to this controversy, namely, theistic evolutionists advocate the complementarity view while young and old earth creationists employ the direct interaction position. To better appreciate this division, it is important to grasp the distinction between primary and secondary causal actions by God. For example, what God did in parting the Red Sea was a primary causal act—an extraordinary, direct, discontinuous intervention by God. What God did in guiding and sustaining that sea before and after its parting involved secondary causal acts by God. In other words, secondary causal acts are God's usual way of operating by which he sustains natural processes in existence and employs them as intermediate agents to accomplish some purpose.

The complementarity view is especially helpful when God acts via secondary causes, for example, in making water from the union of hydrogen and oxygen. The real question is this: In the "natural" history of the cosmos—from its beginning through the origin of life, to the development of various life-forms, including human beings—is there adequate theological, philosophical, and/or scientific reasons for thinking that God exercised primary causality in such a way that science can detect evidence of intelligent design and divine primary causal activity in key episodes (e.g., the origin of life) or in important aspects of living things (e.g., information in DNA or biological complexity)? Complementarians say no and advocate theistic evolution. Old and young earth advocates say yes and adopt what is called "special creation" to express this commitment.

IS CHRISTIANITY A KNOWLEDGE TRADITION?

Lurking in the background of the creation-evolution dialogue is a very important cultural fact that may be more important than the specifics of that discussion: No established, widely recognized body of ethical or religious knowledge exists in the institutions of knowledge in our culture (e.g., the universities). Indeed, ethical and religious claims are frequently placed into some special realm of blind faith or unreality, especially when compared to the authority given to science to define the limits of knowledge and reality in those same institutions.

Some have argued that the rise of modern science has contributed to the loss of intellectual authority in fields such as ethics and religion that are supposedly not subject to the types of testing and experimentation employed in science. Rightly or wrongly, science has been perceived as a threat to the intellectual credibility of Christianity in at least three ways:

1. Some scientific claims call into question certain interpretations of biblical texts (e.g., Genesis 1–2) or certain theological beliefs (e.g., that humans have souls or are made in the image of God).
2. Some scientific claims, if correct, demote certain arguments for the existence of God (e.g., if natural, evolutionary processes can explain the origin or development of life, then we do not "need" to postulate a Creator/Designer to explain these things). There may be other reasons for believing in God, but the advances of science have robbed Christians of a number of arguments that used to be effective.
3. The progress of science vis-à-vis other disciplines like philosophy or theology justifies scientism. In other words, either science and science alone offers true, justified beliefs (strong scientism), or the degree of certainty in science vastly outweighs what these other fields offer (weak scientism), despite the fact that other fields may offer true, justified beliefs.

As evolutionary naturalist George Gaylord Simpson put it:

> There is neither need nor excuse for postulation of nonmaterial intervention in the origin of life, the rise of man, or any other part of the long history of the material cosmos. Yet the origin of that cosmos and the causal principles of its history remain unexplained and inaccessible to science. Here is hidden the First Cause sought by theology and philosophy. The First Cause is not known and I suspect it will never be known to living man. We may, if

> we are so inclined, worship it in our own ways, but we certainly do not comprehend it.[5]

Now Christians must respond to these three problem areas, especially the third one, and the creation-evolution dialogue should be viewed in light of this background. While generalizations can be inaccurate, there are discernible trends that characterize different responses to evolutionary theory. Many theistic evolutionists tend to be suspicious of natural theology (knowledge of the existence and nature of God from features of the created world), especially the design argument for God's existence. Moreover, for many (but not all) theistic evolutionists, science appears to carry more epistemological authority than theology. For example, in his book on theology and science, Arthur Peacocke says, "The aim of this work is to rethink our 'religious' conceptualizations in the light of the perspectives on the world afforded by the sciences...."[6] Elsewhere, Peacocke adds:

> There is a strong prima facie case for re-examining the claimed cognitive content of Christian theology in the light of the new knowledge derivable from the sciences.... If such an exercise is not continually undertaken theology will operate in a cultural ghetto quite cut off from most of those in Western cultures who have good grounds for thinking that science describes what is going on in the processes of the world at all levels. The turbulent history of the relation of science and theology bears witness to the impossibility of theology seeking a peaceful haven, protected from the sciences of its times, if it is going to be believable.[7]

Along similar lines, Karl Giberson claims that "science, after all, is but one limited perspective on the world, although I would argue that it is the most epistemologically secure perspective that we have."[8] Theistic evolutionists with this viewpoint run the risk

---

[5]George Gaylord Simpson, *The Meaning of Evolution* (New York: Bantam Books, 1971), 252.

[6]Arthur Peacocke, *Theology for a Scientific Age* (Minneapolis: Fortress, 1993), 3.

[7]Ibid., 6–7.

[8]Karl Giberson, "Intelligent Design on Trial—A Review Essay," *Christian Scholar's Review* 24 (May 1995): 469; see also *Worlds Apart* (Kansas City: Beacon Hill, 1993); J. P. Moreland, "Theistic Science and the Christian Scholar: A Response to Giberson," *Christian Scholar's Review* 24 (May 1995): 472–78.

of marginalizing belief in God or other theological propositions by turning them into confessional items to be accepted by a mere act of faith without significant rational justification. On this view, the goals of science-theology dialogue are to prevent the evangelical church from being embarrassed by science and to allow scientists the freedom to accept what the majority of their colleagues believe and still consider the claims of Christianity.

These goals are clearly important, but theistic evolutionists face a certain risk in adopting this approach to science-theology and creation-evolution dialogues. They risk revising biblical teaching beyond what is permissible and they raise the suspicion that theological beliefs are private subjective opinions without objective rational justification. Special creationists of both forms are weary of what they perceive to be the constant revising of theology beyond what is permissible biblically or required intellectually. They want to see more cases where theologians can ask scientists to revise their claims in light of well-established theological assertions. They are tired of the "dialogue" always being a monologue in which theologians must constantly adjust to what scientists claim and, eventually, give tacit approval to weak scientism by the constant employment of this complementarity approach.

Fortunately, not all theistic evolutionists agree with this overall approach to the epistemological authority of theology. Some theistic evolutionists offer a number of different views about how we justify belief in God and the Scriptures. Namely, we must simply presuppose God's existence, that belief in God is properly basic (i.e., rationally justified without the need for evidence), the testimony of the Holy Spirit, and some limited role for natural theology. If theistic evolutionists are going to get a broader hearing, they must explain more carefully the following: (1) why they think it is rational to believe in God and in the authority of Scripture; (2) where science needs to readjust its claims in light of theological assertions to avoid the suspicion that it is only theologians who need to adjust and not scientists; and (3) the positive biblical, theological, and philosophical reasons for why theistic evolution should be adopted to prevent it from being viewed as merely a fallback position for those with a low view of the epistemological authority of theology vis-à-vis science. In his essay, Howard J. Van Till does a fine job of addressing these concerns and he offers a promising line of approach for theistic evolutionists who want to gain a hearing in the dialogue process.

In sum, the fundamental issue in the creation-evolution controversy is this: How much of the acceptance of evolution—theistic or naturalistic—is merely the result of accepting scientism and materialistic, naturalistic philosophical assumptions and how much is really justified by the empirical evidence and relevant arguments? If the rules of science (e.g., methodological naturalism) and the scientific evidence go in different directions, with the rules requiring macroevolution and the evidence being best explained by special creation forms of intelligent design, which should we follow? For many young or old earth creationists, evolution is basically applied materialist philosophy and theistic evolution is an unfortunate compromise not required by the evidence. This compromise allows God to play a role in the development of life as long as God's contribution remains entirely invisible, scientifically undetectable, and indistinguishable from the claim that God is just an idea in people's minds. Since issues in the origin of life are not religiously or methodologically neutral (for one's views on this question have profound implications for one's entire worldview), the theistic evolutionary compromise is religiously dangerous, or so say many special creationists.

By contrast, special creationists of both forms tend to see a central role for natural theology and the design argument in justifying theistic belief. They tend to be more concerned about refuting certain widely accepted evolutionary beliefs than avoiding embarrassment in believing what is judged passé by many. As positive as this approach is, special creationism has its own problems as well. Some special creationists seem to adopt an inconsistent approach to the authority of science, often accepting scientific claims in the area of medical science but rejecting what the majority of competent scientists believe about evolution. They also have a tendency to be too suspicious that there are other plausible ways of interpreting the relevant biblical texts apart from their own and they sometimes communicate that they see too limited a role for scientific knowledge vis-à-vis theology in forming a Christian worldview.

In order to get a broader hearing, special creationists must do three things: (1) they must become more involved in standard scientific professional societies and publish more in standard scientific journals; (2) they must explain cases or principles where scientific assertions would cause them to revise their understanding of Scripture; and (3) they must develop the positive merits of creationism as a scientific research program instead of spending so much time offering negative critiques of evolution. In their essays, special creationists Robert Newman (old earth) and Paul Nelson and John Mark Reynolds (young earth) address some of these concerns. It is interesting to note that many special creationists in both camps are shifting the issue from creation versus evolution to intelligent design versus the "blind watchmaker" thesis (the view that blind, mindless Darwinian mechanisms are adequate to explain the origin and development of living things and their parts). On this view, while the issue of the age of the earth is important, the much more pressing problem is to choose the best explanation for the origin of living things given these two noncomplementary alternatives—either by naturalistic mechanisms (with the addition of a scientifically invisible role for a creator for many theistic evolutionists) or by the actions of a designer who does not always employ naturalistic mechanisms in his work as creator and designer. These special creationists claim that we can talk about the age question later and have a good discussion about it once this more pressing issue is settled, namely, the issue of opening minds by legitimating the intelligent design question and showing that the issue has not been settled in favor of theistic or naturalistic evolution as many think.

CENTRAL ISSUES IN THE CREATION-EVOLUTION DIALOGUE

The central issues in the creation-evolution dialogue are philosophical, biblical-theological, and scientific, which we will now survey in the following pages.

1. Philosophical Issues

Perhaps the main philosophical issue in the creation-evolution dialogue involves the relationship between science and theology, especially the debate between advocates of what have come to be called theistic science and methodological naturalism.[9] There has been some controversy as to which field is the proper place to turn in order to seek professional expertise in resolving this debate. Nor is the question of professional expertise merely an academic matter of turf protection, because presently scientists and science educators are the key gatekeepers for the public schools in this area.[10] That a controversy exists can be seen from the following statement by J. W. Haas, Jr., editor of the influential *Perspectives on Science and Christian Faith:* "The place of the philosopher in the practice of science has long been controversial. Whether philosophers should (can?) be the arbiters of what constitutes science remains problematic for the working scientist."[11] Along similar lines, scientist Karl Giberson rejects "the traditional viewpoint that practicing scientists find so annoying, namely that philosophers are the relevant, competent and final authorities to determine the rules of science."[12]

Actually, the issue here is not controversial at all, since the central topics do not involve how to *practice* science (which requires familiarity with instrumentation, procedures, etc.), but how to *define* science and *distinguish* it from nonscience or pseudoscience. To understand this debate and the proper field of study for resolving it, we must first make a distinction between

---

[9]See Howard J. Van Till, Robert E. Snow, John H. Stek, and Davis A. Young, *Portraits of Creation* (Grand Rapids: Eerdmans, 1990); Paul de Vries, "Naturalism in the Natural Sciences: A Christian Perspective," *Christian Scholar's Review* 15 (1986): 388–96; Howard J. Van Till, Davis A. Young, and Clarence Menninga, *Science Held Hostage* (Downers Grove, Ill.: InterVarsity Press, 1988); J. P. Moreland, ed., *The Creation Hypothesis* (Downers Grove, Ill.: InterVarsity Press, 1994), chs. 1, 2; *Christianity and the Nature of Science* (Grand Rapids: Baker, 1989), chs. 1, 6; "Scientific Creationism, Science, and Conceptual Problems," *Perspectives on Science and Christian Faith* 46 (March 1994): 2–13; "Complementarity, Agency Theory, and the God of the Gaps," *Perspectives on Science and Christian Faith* 49 (March 1997): 2–14; Bob O'Conner, "Science on Trial: Exploring the Rationality of Methodological Naturalism," *Perspectives on Science and Christian Faith* 49 (March 1997): 15–30. For a lively debate among Alvin Plantinga, Howard Van Till, Pattle Pun, and Ernan McMullin about the scientific status of creationist theories, see William Hasker, ed., *Christian Scholar's Review* 21 (September 1991).

[10]See Mark Hartwig and P. A. Nelson, *Invitation to Conflict* (Colorado Springs: Access Research Network, 1992).

[11]J. W. Haas, Jr., "Putting Things Into Perspective," *Perspectives on Science and Christian Faith* 46 (March 1994): 1.

[12]Karl Giberson, "Intelligent Design on Trial—A Review Essay," *Christian Scholar's Review* 24 (May 1995): 460.

a first- and second-order issue. A first-order issue is a topic of science about some phenomenon, for example, how to predict earthquakes or manipulate chemical reaction rates. A second-order issue is a topic of philosophy about science itself, for example, its methods, its nature, its differences from other fields. Now the question of how to define science is clearly a topic for philosophers and historians of science. This is not to say that scientists and others cannot be a part of this discussion; it is merely to affirm that when they participate, they will be largely dealing with philosophical issues for which they are not professionally trained.

The fact that these issues are philosophical and not primarily scientific can be seen in the following. Read the relevant debates and discussions and ask what scientific experiment, what scientific procedure one would use to resolve the dispute. Or get any college catalog and look at the course descriptions in different branches of science. You will discover that hardly any undergraduate or graduate program in any branch of science discusses the relevant topics except perhaps during the first week of freshman chemistry. By contrast, entire graduate study programs in the history or philosophy of science are devoted to definitions of science and to drawing lines of demarcation between science and other fields.

What exactly is the debate about the legitimacy of theistic science? Advocates of theistic science hold to these beliefs:

1. God, conceived of as a personal, transcendent agent of great power and intelligence, has through direct, immediate, primary agency and indirect, mediate, secondary causation created and designed the world for a purpose and has directly acted through immediate, primary agency in the course of its development at various times, including prehistory (i.e., history prior to the arrival of human beings).
2. The commitment expressed in proposition #1 above can appropriately enter into the very fabric of the practice of science and the utilization of scientific methodology.
3. One way this commitment can appropriately enter into the practice of science is through various uses in scientific methodology of gaps in the natural world that are essential features of direct, immediate, primary divine

> agency properly understood. When God acts as a primary cause, a gap will be present in the natural world because the effect of his action is a result of his direct causal power and not the result of his guidance of natural processes alone.

In its broadest sense, theistic science is rooted in the idea that Christians ought to consult all they know or have reason to believe when forming and testing hypotheses, when explaining things in science, and when evaluating the plausibility of various scientific hypotheses. Among the things they should consult are propositions of theology. Theistic science can be considered a research program (i.e., a series of theories that exist through time and that are united in some way) that, among other things, expresses a commitment to several ideas—namely, propositions 1 and 2 mentioned on page 18.

A number of Christian scholars reject theistic science and advocate what is sometimes called methodological naturalism, which is basically the idea that theological concepts like God or direct acts of God are not properly part of natural science. Thus, theistic science is fundamentally misguided because it has a faulty philosophy of science and an improper view of how science and theology should be integrated. On this view, the goal of science is to explain contingent natural phenomena strictly in terms of other contingent natural phenomena. Scientific explanations refer only to natural objects and events and not to the personal choices and actions of human or divine agents. Questions about transcendent issues (e.g., ultimate origins, which require a consideration of beings or agents that transcend the physical universe, and the governance of the universe) lie outside the domain of natural science. Advocates of this perspective distinguish methodological naturalism (scientific methodology requires explanation in terms of natural phenomena with no reference to divine action or intention) from metaphysical naturalism (the view that the natural world is all there is) and accept the former but reject the latter.

Does a proper definition of science *require* the adoption of methodological naturalism and the rejection of theistic science as pseudoscience? Remember, this is a second- and not a first-order question. The issue is whether theistic science can count as a scientific research program in the first place, not whether specific theistic science theories currently offer the best scientific models vis-à-vis alternatives. Attempts have been made to show that theistic science is pseudoscience. They take the form of drawing a line of demarcation between science and nonscience (e.g., literature) or pseudoscience (e.g., astrology), a set of necessary and sufficient conditions that some theory, explanation, or research activity must embody to count as science, and of going on to show that theistic science falls on the wrong side of the line of demarcation.

Suggested criteria for what counts as part of the necessary and sufficient conditions for science vary but usually include things like this: the item in question must be falsifiable, be guided by or explained by reference to natural law, be held tentatively, be testable against the empirical world, employ measurable factors, make predictions, be repeatable, be fruitful in guiding future research, and so on.

Space forbids us to investigate these issues further, except to make one point. Historians and philosophers of science are almost universally agreed that theistic science is science and cannot be ruled out as such by demarcationist criteria. Why? Theistic science has been regarded as science throughout most of the history of science, and there are no adequate grounds for thinking that this was wrong. Every line of demarcation used to show that theistic science is pseudoscience has been shown to be neither necessary (some examples of science fail to measure up to the criteria) nor sufficient (there are examples of nonscience that do measure up to the criteria). So even if specific theistic science models are false, empirically inadequate, or in other ways inferior to rival models of the origin and development of life, these models are at least scientific in spite of popular opinions to the contrary. Or so say most experts in the history and philosophy of science.[13]

---

[13]See Larry Laudan, "Commentary: Science at the Bar—Causes for Concern," *Science, Technology, and Human Values* 7 (Fall 1982): 16–19; Michael Ruse, "Response to the Commentary: *Pro Judice*," *Science, Technology, and Human Values* 8 (Winter 1983): 19–23; Larry Laudan, "More on Creationism," *Science, Technology, and Human Values* 8 (Winter 1983): 36–38; "The Demise of the Demarcation Problem," in *Physics, Philosophy, and Psychoanalysis*, ed. R. S. Cohen and L. Laudan (Dordrecht, Holland: D. Reidel, 1982), 111–28; Philip Quinn, "The Philosopher of Science as Expert Witness," in *Science and Reality*, ed. James T. Cushing, C. F. Delaney, and Gary Gutting (Notre Dame, Ind.: University of Notre Dame Press, 1984), 32–53.

2. Biblical-Theological Issues

All the writers in this volume are traditional Christians. They care deeply about theology and the Bible. Everyone wants to have God on his or her side of the debate! They all agree that God is the Creator, that he plays an active role in his creation, and that philosophical naturalism is, therefore, false. There the agreement frequently stops.

For many people the central theological issue of the debate is biblical. What does a fair reading of the Bible most reasonably require of the believer? More importantly, how should key passages of the Bible dealing with creation be read? The status of the biblical record and hermeneutical concerns are central to the debate.

Some Christians think that the Bible provides little or no scientific evidence regarding the origin of life or the universe. The accounts in Genesis 1–11 are not historical. They are written in a literary form that does not demand accuracy in historical details. This position is compatible with an error-free Scripture, since no one demands that the Bible display accuracy beyond its intent. For example, no reader demands "literal" truth of the "scientific" statements in a poem.

By contrast, some Christians think that the biblical accounts in Genesis 1–11 are at least partially historical in nature. For example, they argue that denying a "literal" Adam and Eve has important theological consequences. Moreover, like most traditional readers of the accounts, Jesus himself seems to have regarded Noah as historical. These readers may disagree, as young earth and old earth creationists do, about the nature of the accounts, but they unite in seeing historical importance in them. The "old earth" creationists may allow for large gaps in the genealogical records not acceptable to the "young earth" creationist, but views them as having historical merit. By contrast, some theistic evolutionists deny a real, historical Adam and Eve while others do not.

Young earth and old earth creationists agree on a great deal. They are divided, however, on the theological implications of a second issue: animal death before the fall of Adam. Young earth creationists view this as theologically devastating. How could animals die before the sin of Adam? How could a good God intentionally allow the wasteful and horrific deaths of billions of animals over billions of years before man's sin (much less directly will to employ this struggle as his primary means of creating)? Old earth creationists respond in a number of ways (e.g., animal death before the fall of Adam and Eve was the result of a prior angelic fall; spiritual death resulted from the human fall), but this presents a difficulty for their view.

Another theological problem is often called the "God of the gaps." Proponents of the gaps argument claim that past attempts to find God's "hand" in nature have led to apologetic disaster. Placing God's actions in the gaps in current scientific knowledge restricts his action to those places where science has no explanation for an event or object. This lack of knowledge is the gap into which the Christian thinker tries to fit God. By placing God in the gap, this sort of theist gains a short-term apologetic advantage. The critic of the gap approach argues, however, that in the long term such a policy is misguided. Eventually science discerns a natural explanation for the phenomenon in question, the putative gap disappears, and theism is damaged. What is more important, the gaps in human understanding of the world are growing smaller and smaller. If Christians pursue the gap argument, the place of God in the natural order of things will soon disappear. Many contemporary Christian writers, particularly Christian scientists, take this God-of-the-gaps problem quite seriously.

On closer examination, the gaps argument turns out not to be an actual argument. It is more a bit of apologetic advice. Few contemporary philosophers have risen to defend this worry as being legitimate. Despite the claims of worried Christian scientists, the God-of-the-gaps strategy has had limited historical applicability, at least in the contemporary setting. There are few, if any, cases of serious Christian thinkers actually falling into gap thinking. Merely postulating the action of an agent is not an appeal to a gap. Otherwise, every criminal detective who believes an agent committed a crime is guilty of "giving up" on a more "naturalistic" explanation!

The gaps argument has also turned out to be bad counsel on three grounds: (1) it makes the false assumption that theories about God's actions are rendered false by the mere existence of (sometimes quite implausible) naturalistic accounts dealing with the same events; (2) it fails to define any of its terms. It is, therefore, too vague to be of any practical value. For example, what counts as a gap in human knowledge? What if science itself discovers a limit to scientific investigation? This would not be a gap waiting to be filled but rather a boundary. What is the difference between a gap and a boundary of science? And (3) it does not adequately represent theories like intelligent design, which are not motivated by a gap in human understanding. Most old and young earth creationists who appeal to divine, miraculous actions to explain the creation of some entity do so because they think there are adequate positive, scientific, philosophical, or theological reasons to justify this move. They do not "appeal to God" merely to cover the ignorance of a supposed naturalistic mechanism, nor do they limit God's activity in nature to cases of primary causes. They think that we are regularly confronted with human artifacts that result from human intelligent design and that we have observational evidence to claim that various features of the biological world bear the characteristics of intelligent design and, therefore, the best explanation of the positive evidence is the design hypothesis.

In the final analysis, many think that what lies behind the God-of-the-gaps complaint is the airtight, rigid employment of methodological naturalism. These respondents to the gaps argument claim that the methodological naturalism in the so-called rules of science are at odds with the scientific evidence which, in certain cases, supports the claim that the primary causal activity of God as Creator/Designer is the best explanation of the relevant scientific evidence. Thus, the gaps argument has the effect of closing minds to the evidence in the interest of following a set of rules that inadvertently express and, in turn, support scientism and philosophical naturalism, regardless of how much theistic advocates of the gap argument desire to allow God a scientifically undetectable role in creation.

3. Scientific Issues

We have little space to outline all the scientific issues that are subject to debate in this controversy. An important fact to keep in mind is that people on all sides of the controversy frequently agree about the brute evidence. They do not, however, agree on how to interpret that evidence. This is evident in all of the main articles in this volume. All of the writers focus their articles on how to see the evidence but spend little or no time arguing over the evidence itself. This is to be expected. Short of the few irresponsible popularizers, who often misrepresent the state of the evidence, there is little question about the "facts." For example, almost all responsible thinkers agree that certain chemical elements now decay at known rates. There is little disagreement about this. So what should be made of these facts and how should they be interpreted? On this there is *widespread* disagreement amongst Christians!

The main scientific divide is between those who think that the scientific evidence for Darwinian evolution is conclusive and those who think it is not. Leaving aside philosophical arguments for Darwinism (e.g., a good, wise God would not design a world with imperfections or with animals with such effective means of inflicting pain and death on other organisms), young and old earth creationists think that it is reasonable to doubt the truth of Darwinism. Some would go even further and suggest that belief in Darwinism is not reasonable at all without the support of philosophical naturalism. On the other hand, theistic evolutionists deny that they are naturalists but are still satisfied that present evidence strongly favors evolution. Some of them would argue that denying the truth of evolution is itself evidence of an irrational religious commitment.

Much hinges on what is meant by *biological evolution*, which can be understood in at least four different ways. First, evolution is sometimes referred to as a "fact." For example, it is a fact that species and other life-forms have changed over time. Finch beaks get smaller and shorter. Dogs can be bred to be bigger or smaller. If evolution simply means "change over time," then all of the writers in this book believe in evolution. "Microevolution" of the sort easily observed by any thoughtful human being is conceded by even the most hard-core critic of Darwin.

Second, evolution can also be used to refer to "macroevolutionary change." This sense of evolution allows for naturally occurring change all the way up to the phyla and kingdom level of biological organization. Usually, proponents argue that observations such as those related to the fossil record and studies in biochemical similarities strongly favor this possibility. On this view, given time and the right mechanism, biological change can indeed be radical. This view is usually associated, though it need not be, with the idea of common descent. All organisms share a common biological ancestor, that is, there is one "tree" of life to which every living thing is connected. These ideas are rejected by almost all old and young earth creationists. They are accepted both by theistic evolutionists and the scientific mainstream.

A third way of understanding evolution is as a shorthand reference to the neo-Darwinian mechanism to produce "change over time." Natural selection, random mutations, time, chance, and other mechanisms produce all the life-forms now in existence. There is no need to appeal to direct divine activity. Things around humankind may appear to be designed, but they were not. They are the product of a naturalistic "blind watchmaker." They appear to be designed, but are merely adapted for their environment. This is what most mainstream scientists mean by evolution.

Many theistic evolutionists would blanch at the naturalistic implications of the blind watchmaker hypothesis. They would allow for some form of divine providence to provide purpose and design to the process of evolution, even granted the truth and the effectiveness of the neo-Darwinian mechanism. The difficulty is that while some divine action and purpose may be logically compatible with the blind watchmaker, many contemporary intellectuals see no need to appeal to divine action and purpose at all, given the adequacy of naturalistic mechanisms for explaining the origin of living things and their parts. As Phillip E. Johnson has pointed out,

> Politically astute scientific naturalists feel no hostility toward those religious leaders who implicitly accept the key naturalistic doctrine that supernatural powers do not actually affect the course of nature.... The most sophisticated naturalists realize that it is better just to say that statements about God are "religious" and hence incapable of being more than expressions of subjective feeling. It would be pretty ridiculous, after all, to make a big deal out of proving that Zeus and Apollo do not really exist.[14]

Elsewhere, Johnson observes,

> The conflict between the naturalistic worldview and the Christian supernaturalistic worldview goes all the way down. It cannot be papered over by superficial compromises.... It cannot be mitigated by reading the Bible

---

[14]Phillip E. Johnson, *Defeating Darwinism* (Downers Grove, Ill.: InterVarsity Press, 1997), 100–101.

> figuratively rather than literally.... There is no satisfactory way to bring two such fundamentally different stories together, although various bogus intellectual systems offer a superficial compromise to those who are willing to overlook a logical contradiction or two. A clear thinker simply has to go one way or another.[15]

Johnson's remarks serve as a reminder that theistic evolutionists have to offer reasons why they believe in Christian theism in the first place in order to avoid achieving a "reconciliation" between Darwinism and Christian theism at the price of placing the epistemological authority of Christianity in some private, upper story.

Finally, many leading scientists such as Richard Dawkins and Carl Sagan have used evolution in a philosophical way. They use it as a shorthand for scientific naturalism. This is an illegitimate use of a scientific term.

What is the status of the evidence? How secure is the blind watchmaker hypothesis based on the "facts" of science? That is the scientific portion of this debate.

HISTORICAL OVERVIEW OF THE ISSUE

Dogmatic opinions about the origin and formation of life are all too common in contemporary discussions in philosophy, science, and theology. This unfortunate state of affairs is understandable in light of the importance of the issues involved. Every culture clings to its particular creation myth. The question of the relationship between a creator and his creation has a long and complex history. For most of the history of Western thought, the idea of "design" or a creator acting in space and time was perfectly acceptable in rational discourse. In the last two hundred years, this notion has come under increasing attack.

As we will see, a conviction exists in the contemporary academic community that the criticisms of design in the modern world, beginning with Hume and culminating in Darwin, have indeed won the field. The argument from design is dead. Design in the natural world itself is best explained away on Darwinian

---

[15]Ibid., 111.

grounds as "apparent design" that is best understood in naturalistic terms.

In the space of this brief introduction, we barely have time to trace more than the bold outlines of the situation. We can, however, trace a "bird's-eye view" of the rise and fall of the idea of design and creation in Western thought.

As is usually the case, the discussion began with the ancient Greeks. Plato made the importance of design and creation quite clear in his last and longest dialogue, *The Laws* (Book X). He said:

> *Athenian:* Quite true, Megillus and Cleinias, but I am afraid that we have unconsciously lighted on a strange doctrine.
>
> *Cleinias:* What doctrine do you mean?
>
> *Ath.* The wisest of all doctrines, in the opinion of many.
>
> *Cle.* I wish that you would speak plainer.
>
> *Ath.* The doctrine that all things do become, have become, and will become, some by nature, some by art, and some by chance.
>
> *Cle.* Is not that true?
>
> *Ath.* Well, philosophers are probably right; at any rate we may as well follow in their track, and examine what is the meaning of them and their disciples.
>
> *Cle.* By all means.
>
> *Ath.* They say that the greatest and fairest things are the work of nature and of chance, the lesser of art, which, receiving from nature the greater and primeval creations, molds and fashions all those lesser works which are generally termed artificial.
>
> *Cle.* How is that?
>
> *Ath.* I will explain my meaning still more clearly. They say that fire and water, and earth and air, all exist by nature and chance, and none of them by art, and that as to the bodies which come next in order—earth, and sun, and moon, and stars—they have been created by means of these absolutely inanimate existences. The elements are severally moved by chance and some inherent force according to certain affinities among them—of hot with cold, or of dry with moist, or of soft with hard, and

> according to all the other accidental admixtures of opposites which have been formed by necessity. After this fashion and in this manner the whole heaven has been created, and all that is in the heaven, as well as animals and all plants, and all the seasons come from these elements, not by the action of mind, as they say, or of any God, or from art, but as I was saying, by nature and chance only. Art sprang up afterwards and out of these, mortal and of mortal birth, and produced in play certain images and very partial imitations of the truth, having an affinity to one another, such as music and painting create and their companion arts. And there are other arts which have a serious purpose, and these co-operate with nature, such, for example, as medicine, and husbandry, and gymnastic. And they say that politics cooperate with nature, but in a less degree, and have more of art; also that legislation is entirely a work of art, and is based on assumptions which are not true.
>
> *Cle.* How do you mean?
>
> *Ath.* In the first place, my dear friend, these people would say that the gods exist not by nature, but by art, and by the laws of states, which are different in different places, according to the agreement of those who make them; and that the honorable is one thing by nature and another thing by law, and that the principles of justice have no existence at all in nature, but that mankind are always disputing about them and altering them; and that the alterations which are made by art and by law have no basis in nature, but are of authority for the moment and at the time at which they are made.—These, my friends, are the sayings of wise men, poets and prose writers, which find a way into the minds of youth.

Plato described the philosophy of his day that attributed the creation and design of the universe to chance. He suggested that this theory places the origins of the divine and moral norms in an "art," which is itself the product of chance. There is no design or art behind the cosmos. In short, the rational and the orderly sprang from the irrational and disorderly by the workings of chance and some scientific process. Plato, or at least the Athenian of the dialogue, strongly opposed teaching this view in his mythical city-state of Magnesia. He feared it would harm the minds and morals of the youth of the city. Many contemporary arguments are prefigured in this early discussion.

Plato presented the two basic options in his dialogues. The universe might have been created under the control of a craftsman, the Demiurge of the Timaeus. In this case, Mind came before cosmos. Of course, it needed not to be the case that Mind was paramount, as in the Christian conception. Plato's craftsman had to work with recalcitrant matter. This god could only do his best in creation. It still was the case, however, that the cosmos had a fundamental basis in the rational workings of a divine Mind. On the other hand, there is the position we have just examined in the *Laws*. To paraphrase Alfred North Whitehead, most debates about the design or creation of the cosmos are a footnote to this discussion in Plato.

Both Plato and Aristotle saw knowledge of Mind as fundamental to gaining knowledge about the physical cosmos. On the other hand, the philosophies of the Epicureans and Lucretius demanded no god for their cosmological accounts. The Hellenistic and Roman periods were a time of lively discussion in this area. It is safe to say, however, that forms of neo-Platonism and Stoicism (both theistic) came to dominate the later periods of intellectual thought. The apparent order of the universe, and perhaps the need for order in the Roman state, made the more theistic design position the dominant one.

The spread of Christian theism in the Roman Empire encouraged this process. The apologists for the new faith were able to rely in their works on the widespread acceptance of the need for a creator. For example, the second-century apologist Theophilus of Antioch was able to rely on such beliefs in his writing. In the twilight of the western Roman Empire, Augustine was able to appropriate the same sort of inclination toward the reality of design in his *City of God* and his discussion of creation in the latter portions of his *Confessions*.

In the medieval West, design and the involvement of supernatural intelligence were widely accepted, and they helped justify the idea that we live in a creation that can be studied and in which truths can be grasped beyond the surface appearances of things. Contrary to the stereotype of the period as a time of intellectual stagnation and dogmatism, philosophy of science continued to develop during the Middle Ages. Such men as Roger

Bacon, Duns Scotus, and William of Ockham made important advances of the Greco-Roman understandings of the natural world and philosophy.

These general inclinations strongly in favor of design did not change with the advent of the modern age. Seeing the work of an intelligent designer was a commonplace for the early scientists. Isaac Newton, who ushered in new methods in understanding the natural world, was a lifelong student of the Bible and had no difficulty seeing strong evidence for design in the universe. Even the most severe critics of the Christian religion in the period of the Enlightenment, which helped birth the modern world, did not reject the existence of a creator or providence behind the cosmos. Thomas Paine, the archcritic of traditional Christianity, speaking in the *Age of Reason* about the death of the established church said, "Human inventions and priestcraft would be detected; and man would return to the pure, unmixed and unadulterated belief of one God, and no more." Reason demonstrated the providential order of things to such men. The self-proclaimed "infidels" of the Enlightenment were not for the most part atheists in our modern sense, and the order of nature was a powerful reason for their theism.

Two assaults changed this picture, seemingly forever. First, David Hume launched a philosophic assault on the "argument from design." He said:

> You then, who are my accusers have acknowledged, that the chief or sole argument for a divine existence (which I never questioned) is derived from the order of nature; where there appear such marks of intelligence and design, that you think it extravagant to assign for its cause, either chance, or the blind and unguided force of matter. You allow, that this is an argument drawn from effects to causes. From the order of the work, you infer, that there must have been project and forethought in the workman. If you cannot make out this point, you allow, that your conclusion fails; and you pretend not to establish the conclusion in a greater latitude that the phenomena of nature will justify. These are your concessions. I desire you to mark the consequences.
>
> When we infer any particular cause from an effect, we must proportion the one to the other, and can never

> be allowed to ascribe to the cause any qualities, but what are exactly sufficient to produce the effect. A body of ten ounces raised in any scale may serve as a proof, that the counterbalancing weight exceeds ten ounces; but can never afford a reason that it exceeds a hundred. If the cause, assigned for any effect, be not sufficient to produce it, we must either reject that cause, or add to it such qualities as will give it a just proportion to the effect.[16]

Hume went on to argue that one could never derive a knowledge of a personal, intelligent God from the impersonal cosmos. Any argument from design or to design was not robust enough to do the work many theists had traditionally assigned to it, that is, the establishment of the existence of the full-blown God of Christian theism. The complexity so evident in the natural world, however, still led most thinkers to favor a creationist position.

Darwinism changed the picture substantially. Before Darwin, theists could point to natural objects like the eye and then challenge their philosophically inclined critics to provide a better explanation than theism. Darwin provided a purely naturalistic account for apparent design in the natural world. In the *Origin of Species* he challenges his critics, "It is so easy to hide our ignorance under such expressions as the 'plan of creation' or 'unity of design,' etc., and to think that we give an explanation when we only restate a fact."[17] Darwin would have none of that kind of "sloppy thinking." Instead, he proposed a mechanism—natural selection—that would do the work of providing for the patterns in nature that others had only passively described.

The results of the debate over design in nature for theism were very great. In the words of the contemporary defender of neo-Darwinism, Richard Dawkins, "Darwin made it possible to be an intellectually fulfilled atheist." One can see the impact of Darwin on the argument from design in the writings of the philosopher William James. Writing in *Varieties of Religious Experience* as the twentieth century dawned, he said, "As for the argument from design, see how Darwinian ideas have revolutionized

---

[16]David Hume, *An Enquiry Concerning Human Understanding* (Indianapolis: Hackett, 1993), 93.

[17]Charles Darwin, *Origin of Species* (New York: Signet Books, 1958), 444.

it. Conceived as we now conceive them, as so many fortunate escapes from almost limitless processes of destruction, the benevolent adaptations which we find in Nature suggest a deity very different from the one who figured in the earlier versions of the argument."[18] God is, at best, unemployed in the new cosmology. In light of the putative failure of the design argument, many scientists and philosophers dispensed with him altogether.

The reaction to Darwin and Darwinism within the Christian community was mixed. The first and most vocal critics of Darwin in England were his fellow scientists. Most theologians had been prepared for the idea of some sort of progressive creation for a long time. Darwinism was in the air in the form of poetic and cultural paeans to progress. Earlier findings in geology had challenged a literal reading of the Genesis chronology and a single, global flood in the days of an actual Noah. Many Christians, especially in the universities, were willing to take Genesis 1–11 as less than actual history. At first, therefore, the majority of English theologians reacted carefully and in a guarded manner to Darwinism. In fact, no large-scale creationist response to Darwin ever developed in England, despite well-known and vocal critics such as Prime Minister Gladstone and the noted scientist Lord Kelvin.

The United States was, however, a different story. There existed a much larger population of more conservative Christians. They were less willing to view Darwin and Darwinism with equanimity. Again, many Christians, even the most conservative, had already accepted the age of the earth and the extent of the Flood as an open question. Such conservative theologians as Charles Hodge and B. B. Warfield were open to these ideas, but reacted more strongly to Darwinism. At the very least, Darwin seemed to limit severely any active role for the Creator in the creation process. Hodge summed up the feeling of many when he baldly equated Darwinism with atheism.

The struggle against Darwinism and modernism in the academy, even in the United States, was a losing battle. By the middle of the twentieth century, opposition to Darwinism was limited to the more fundamentalist religious communities. Groups like the Seventh-day Adventists carried on an active assault

---

[18]William James, *Varieties of Religious Experience.* Library of the Future Series, screen 711.

against evolutionary thinking, sometimes with more noise and vigor than scientific care or rigor. Large numbers of scientists may have been "theistic evolutionists," reserving for God some nearly invisible role in the process of evolution, but their voice was very quiet in the general culture. The public perception was that Darwinism, and some form of naturalism, had triumphed.

Most unexpectedly, however, the critics of Darwinism were reinforced by two important groups at the mid-century point. First, many conservative Christians began to receive a larger number of graduate degrees in science. In the 1940s the American Scientific Affiliation (ASA) brought a small but growing number of evangelicals together who had training in the sciences. At the very least, this group tended to be cautious about Darwinism. It allowed both theistic evolutionists and critics to dialogue in a sustained and responsible manner. The ASA created a respectable forum and some peer review to critics of evolutionary thought.

A smaller and more conservative group of scientists, who rapidly became more influential with laypeople in traditionalist Christian circles, developed a more radical critique of the reigning paradigm. They were the "young earth creationists." With the publication of *The Genesis Flood*[19] by Henry Morris and John Whitcomb, the movement came to full flower. They insisted that both scientific and biblical evidence favored an earth, and usually a universe, younger than ten thousand years. The young earth creationists also explained many of the features of the modern world by recourse to Noah's flood. Organizations such as the Creation Research Society, the Bible-Science Association, and Morris's own Institute for Creation Research spread these alternative views. Frequently plagued by irresponsible advocates and sloppy scientific methodology, the movement was widely ridiculed and attacked by the secular media and academy. Their influence tended to be quite limited even in the evangelical and fundamentalist academic mainstream.

By the end of the twentieth century, five important developments changed the landscape of this dialogue. First, the rise of postmodernism in the secular world assaulted the confidence of the advocates of Darwinism from within the academic mainstream. Feminist and other critics of modernity questioned the

---

[19](Philadelphia: Presbyterian and Reformed, 1961).

"truth" of scientific theories. This assault on the left sapped some of the strength of the scientific establishment.

Second, young earth creationism began to develop a broader and more sophisticated line of reasoning. Under the leadership of Morris and other well-trained scientists, the movement saw an ever-increasing rigor to its publications. By the late 1980s and the 1990s, meetings like the quadrennial International Conference on Creationism attracted hundreds of scientists, theologians, and philosophers. More and more journal articles and conferences were more adequately reviewed by peers and were of a much higher quality than earlier efforts. Currently, efforts are being made to present a positive theoretical model and not just to attack evolutionary ideas. While success in this latter area remains elusive, the quality of the effort has shown a geometrical improvement.

The third important shift in the Darwinian debate came with a more thoughtful and theologically conservative form of theistic evolution. Scientists, often within the ASA, developed a new way of integrating the findings of modern science and theology that seemed to allow for God's action in creation without limiting the scope of scientific inquiry.

Fourth, a renewed interest in old earth creationism in academic and lay evangelical circles has been in evidence. These Christians accepted evidence for an old earth and universe, while rejecting Darwinian evolution. They usually argued for a literal Adam and Eve and a local deluge. Often caught in the middle—too accommodating for young earth creationists on biblical issues and too conservative for theistic evolutionists—their penetration into the Christian community had been somewhat limited. The ministry of popular writers and speakers such as Hugh Ross gave their ideas new prominence by the end of the 1990s.

Finally, a new intelligent design movement was sparked by the work of Phillip E. Johnson, a University of California at Berkeley law professor. He challenged Darwinism in his works *Darwin on Trial* and *Reason in the Balance*.[20] The intelligent design movement saw contemporary scientific evidence pointing

---

[20]*Darwin on Trial* (Downers Grove, Ill.: InterVarsity Press, 1991); *Reason in the Balance* (Downers Grove, Ill.: InterVarsity Press, 1995).

toward a "design hypothesis." At the same time, philosophers William Dembski, Paul Nelson, and Stephen Meyer developed a new approach to design and design arguments that did not seem vulnerable to traditional criticisms. This movement has been neutral on biblical questions and embraces both theists and nontheists. It contains active young earth and old earth creationists.

This current volume allows for an active debate from traditional Christians representing these views. Paul Nelson and John Mark Reynolds are among the most responsible of a new generation of young earth creationists who are also involved in the intelligent design movement. Robert Newman represents a resurgent old earth creationism. He is also an important player on intelligent design. The work of Howard Van Till represents the most theologically and scientifically responsible attempt to accommodate evolution to conservative Christian theology. Respondents range across the ideological spectrum.

FOCUS QUESTIONS

In order to get at the most important issues at the core of the creation-evolution dialogue, we have asked a qualified representative of each of the three major positions to respond to five sets of questions listed below.

Set 1—*Overall Position.* What is your position on the creation-evolution controversy? What is your broader view of the integration of science and theology? What role does your view of integration play in determining your position on creation and/or evolution?

Set 2—*Importance of the Topic.* Why does this controversy matter? What do you take to be the broader cultural and intellectual implications that are related to this debate? Describe your own journey in relation to the creation-evolution controversy. As a younger believer, what was your position, and what key factors contributed to maintaining or changing your view?

Set 3—*Philosophy of Science.* How does your understanding of the nature of science, the "scientific method," and the nature of scientific evidence influence your approach to the Bible and how have they shaped your theological beliefs? Put differently, describe your philosophy of science (e.g., scientific methodology) and explain how your philosophy of science informs your approach to the Bible and your theological beliefs.

Set 4—*Theology and Scripture.* How does your understanding of the Scriptures, hermeneutics, and theology affect your views on the nature of science, the relevant scientific data, and your position on the creation-evolution controversy?

Set 5—*Epistemology.* What roles do extrabiblical scientific evidence and arguments in natural theology play in confirming or contesting your theological beliefs? When it comes to resolving an apparent tension between science and theology, which, in general, carries the most epistemological weight? As part of answering these questions, say something about your own epistemology and how it informs your view of the relationship between faith and reason in general, and science and theology in particular.

In addition to these questions, we have asked each of our three advocates to end his chapter by offering advice to Susan, a hypothetical student who has come to that person for advice. Susan, an evangelical Christian raised in a home that taught her to accept an old earth creationist position, has not developed her views carefully. So far in college she has come to learn that the vast majority of scientists accept either naturalistic or theistic evolution and she has heard some good arguments for evolution. If these arguments are so convincing, she wonders, then who would want to be a Christian theist in the first place? Recently, Susan has been made aware of a new movement advocating theistic design, and she has read some of the books advocating this approach. This has tended to reinforce her commitment to progressive creationism. However, last weekend she attended a conference where she heard a case for the young earth view that made her open to this as well. We have asked our three advocates to offer Susan guidance for why she should accept that person's view of the creation-evolution dialogue.

This book brings together a number of respected Christian thinkers with different perspectives and a wide diversity of academic backgrounds. Paul Nelson's area of expertise is in biology and the history of science, and he received training in these areas at the University of Chicago. John Mark Reynolds, a professor of philosophy at Biola University and director of Biola's Torrey Honors Institute, specializes in Greek philosophy and the philosophy of science. Robert C. Newman holds a Ph.D. in theoretical astrophysics and is a professor of New Testament at Biblical Theological Seminary. Lastly, Howard J. Van Till is a professor of physics at Calvin College.

Each chapter will include four responses from different academic disciplines: biblical studies/hermeneutics (Vern S. Poythress, Westminster Theological Seminary); theology (John Jefferson Davis, Gordon-Conwell Theological Seminary); philosophy (J. P. Moreland, Talbot School of Theology, Biola University); and natural science (Walter L. Bradley, Texas A & M). Responding to the dialogue are two additional people from different areas of training: Richard H. Bube, a scientist who is now professor emeritus of materials science and electrical engineering at Stanford University, and Phillip E. Johnson, a professor of law at the University of California at Berkeley.

Since young and old earth creationists have more in common with each other than they share with theistic evolution (e.g., young and old earth creationists advocate the recent intelligent design movement and theistic evolutionists do not), we have, in the interests of fairness, given Van Till a word limit equal to the combined total words assigned to Nelson and Reynolds and Newman. Also, in the final reflections, we have selected one representative of the intelligent design movement (Johnson), who speaks both for young and old earth creationists in this regard, and one representative of theistic evolution (Bube) to provide balance and fairness.

Chapter One

YOUNG EARTH CREATIONISM

Paul Nelson and John Mark Reynolds

YOUNG EARTH CREATIONISM

Paul Nelson and John Mark Reynolds

1. OVERALL POSITION

We hold the view of *recent* or so-called *young earth* creation. Unfortunately, neither "young earth" nor "recent" is satisfactory as a descriptive adjective. If you are asked to give your age on a legal form, you do not write "old," "young," "recent," or any other relative term; rather, you give an exact number. The world is precisely as old or as young as it actually is. Young earth creation is thus a confusing misnomer, seeming to imply that the earth or universe are young relative to some unspecified ("old") temporal reference point. (The *virtuosi* of the early scientific revolution thought that God had created the world within the biblical time span, but they would not therefore have described their cosmology as young or recent. They would have asked, "Young or recent in reference to *what?*") This caveat should always be kept in mind when using terms like "young earth" or "recent," which actually describe, and then imperfectly, differences between various theories about the timing of creation.

The young earth creationism position is that most often identified as "creationism" by the majority of scientists, educators, and the press, largely because (at least until recently) those persons most likely to come to public attention in the creation-evolution controversy as wanting changes in science education, urging legislation, or debating the issue on university and college campuses, were also most likely to hold young earth views. While neither "recent" or "young earth" is entirely satisfactory as a descriptive adjective, both terms can reasonably be applied to the general position we shall now describe.

The main distinguishing features of the recent creation position are:

1. An open philosophy of science. (We define "open" in detail under "3. Philosophy of Science" below.)
2. All basic types of organisms were directly created by God during the creation week of Genesis 1–2.
3. The curse of Genesis 3:14–19 profoundly affected every aspect of the natural economy.
4. The flood of Noah was a historical event, global in extent and effect.

We amplify each of these points at the end of this section. Other distinctive aspects of the recent creation position (e.g., a historical Adam and Eve, directly created by God as the original parents of humankind) follow from these cardinal claims.

The largest organizations advocating recent creation include:

The Creation Research Society (CRS), established in 1963, with the geneticist Walter E. Lammerts as its first president. A scholarly society chartered solely for research and publication, the CRS includes six hundred voting members (defined as such by holding graduate degrees in science) and eleven hundred nonvoting members and publishes the *Creation Research Society Quarterly*, a technical research journal now in its thirty-fourth volume.

The Geoscience Research Institute (GRI), affiliated with Loma Linda University in Loma Linda, California, and Andrews University in Berrien Springs, Michigan. Established in 1958, the GRI, staffed by scientists from the Seventh-day Adventist church, publishes the biannual scholarly journal *Origins*, now in its twenty-fourth volume.

The Institute for Creation Research (ICR), Santee, California, established in 1972 by Henry Morris as an offshoot of Christian Heritage College in San Diego, California. ICR publishes books, technical monographs, and the monthly magazine *Acts and Facts*, and broadcasts an international radio program. Most of the best-known proponents of the young earth position (e.g., Henry Morris, Duane Gish, Ken Ham) are, or were, associated with the ICR.

Organizations and societies advocating recent creationism exist in other countries as well (e.g., Canada, Australia, Germany, England, and Korea), and dozens of local groups operate at the state and regional level, such as the Creation Science Fellowship in Pittsburgh, who sponsor the quadrennial International Conference on Creationism (ICC). Scientists currently working within the recent creation perspective include paleontologist Kurt Wise of Bryan College, geophysicist John Baumgardner of Los Alamos National Laboratory, physicist Russell Humphreys of Sandia National Laboratory, biologist Wayne Frair of the CRS, microbiologist Siegfried Scherer of the Technical University of Munich, and paleoanthropologist Sigrid Hartwig-Scherer of the University of Munich.

Let us now return briefly to the key features of the recent creation position. It may be helpful to consider each in a comparative table:

	Recent	Progressive	Theistic Evolution
Open philosophy of science	+	+	–
Biological discontinuity and design	+	+	–
Curse affecting natural economy	+	–	–
Global flood	+	–	–

It will be generally true that recent and progressive creationists agree that one may infer God has acted directly from patterns of physical evidence (i.e., both schools of thought hold what we call an "open" philosophy of science). And, indeed, both recent and progressive creationists see the evidence as pointing to the direct action of God as a primary cause in the history of life. God created the major divisions of animals and plants directly. Furthermore, both positions hold the living state, life itself, to be a phenomenon that cannot be reduced to or derived from lower-level physical and chemical entities or causes. Organisms were directly designed by God as systems that, while they certainly rely on physical and chemical laws, would never come into being via those laws alone.

For their part, however, theistic evolutionists generally advocate a more restrictive philosophy of science (as we explain below under "3. Philosophy of Science"). Physical objects, including organisms, may be explained only by physical causes. Theistic evolutionists also hold that all organisms share a common ancestor, and many (if not most) also hold that the first organisms were themselves self-replicating systems that arose via physical causes from nonliving things.

Recent and progressive creationists differ, however, about the scope and nature of the curse of Genesis 3:14–19. Progressive creationists tend to view the days of creation as long periods of time, and therefore see animal death and suffering existing long before Adam sinned. But recent and progressive creationists differ most sharply on the extent of the flood of Noah and the details of earth and astronomical history. If one wants a reason not to hold the recent creationist position, one would typically begin here—at the bottom of the table shown above—with geology.

Not coincidentally, we have listed these points in relation to our confidence in discussing them, and our sense of their relative strength and importance. Paul Nelson is trained as a philosopher and biologist. John Mark Reynolds is trained as a philosopher. Thus, in front of any professional audience, we would begin our case for creation with the evidence and arguments we know best, namely, the philosophy of science and biology. In this book, however, we are writing on a far broader canvas, as Christians, to Christians. Put another way, as Christians we would argue from biblical grounds for an absolutely open philosophy of science, even if we doubted or were completely unconvinced of any of the other points. God acts in history, directly, as he pleases, leaving unmistakable evidence of his existence. No other philosophy of nature, and hence, no other philosophy of science, can be reconciled with the biblical worldview. It should surprise no one that the heart of the public controversy has come to center on the philosophy of science.

Given an open philosophy of science, we think the evidence for biological discontinuity and design is unequivocal, indeed, overwhelming. Organisms cannot be understood except as the products of a directly acting, purposeful intelligence. By their very nature, physical laws and regularities are insufficient to generate the information and complexity necessary for life, and within the next few decades this may be experimentally demonstrable.

Moreover, our best understanding of the main patterns of fossil evidence and developmental biology point clearly to deep and original discontinuities in the relationships of organisms. The overall geometry of the history of life really does not depict a single tree of life, as Darwin thought, but a forest of trees, each with its own independent root. If this conclusion were not philosophically distasteful to biologists—that is, if it did not point so clearly to intelligent design—it would, we believe, be adopted without hesitation by the scientific community.

The case for primary discontinuity is powerfully capped by our growing understanding of the natural limits on biological variation. In a sense, the very last thing a multicellular organism wants to do is to evolve, that is, to vary fundamentally in its form and function, as required by evolution. Both unicellular and multicellular organisms are constrained by functional demands, and, in the case of developing organisms (i.e., animals), those requirements strongly limit what variations are tolerated. Darwin did not know this. Contemporary thinkers do. Thus, many moderns no longer seek justification for evolution in biological evidence because it no longer resides there. They go straight to the philosophy of science. The common ancestry of organisms by natural selection flows as a scientific theory not from what we know about life but from the philosophy of naturalism, which binds what scientists may infer about nature.

Some Christians do believe that evolution was God's means of creating the cosmos. This is certainly logically possible. God is free to do whatever he wills. But the question remains, What did he do? There are two lines of evidence available to the Christian: religious (the teachings of the church and Scripture) and natural (the findings of science). In our opinion, both types of evidence point away from evolution.

Why do some good, sincere religious people accept evolution? It is impossible, of course, to get inside the mind of any individual person. There is no denying the sincere religious faith of many of these folk. Many are well-trained, competent scientists. They are rarely leaders of mainstream science, but they are allowed an existence within their disciplines. Theistic evolution is not the result of some stupidity, but a creative failure. Such people, for whatever reason, cannot see beyond the bounds of their training or their own philosophic and theological commitments to seriously consider other possibilities.

Christians who are theistic evolutionists are in a cruel bind. Their theology demands a God who acts in space and time. They are captured, however, by a methodological naturalism in science that will not allow them to scientifically consider positive evidence for a creator. They are so fearful of being wrong about proclaiming God's activity in the natural world that they have decided that his activity is invisible to human science. As we shall see, this limitation of science impedes the ability of theistic evolutionists to consider all the possibilities. It also raises a theological worry. How can theistic evolutionists avoid worshiping an unemployed God?

Theologically conservative Christians who are theistic evolutionists must avoid this sort of deistic creator. How can this be done? They often resort to speaking of a God who "sustains" the universe. Without God's sustaining power, the universe would simply cease to exist. But this clever solution simply hides the problem under theological jargon.

At the very least, God's power causes each bit of the cosmos to continue to "be." Existence, however, is open to scientific study. Physics, for example, can legitimately investigate why an object or a particle exists or continues to exist. If God is at the bottom of it all, and only a naturalistic answer is acceptable to science, then in the end natural science will be left with a gap in its knowledge. Without an ability to turn to the supernatural, science will be left with hopelessly false naturalistic speculations about the reason physical objects exist. Worse still, a prior commitment to methodological naturalism will lead these selfsame scientists to continue looking for a naturalistic basis for existence when there is none. The entire position is doomed to research futility.

This problem can be brought even closer to home. At this very moment, you are engaged in a nonnaturalistic event. Traditional Christianity teaches that your nonphysical soul is engaged with your body in the task of reading. You are, if science must be naturalistic, engaged in an activity that science will never understand. Science bound by naturalism will never be able to recognize an immaterial soul. Reading is not scientifically explainable. This holds true for whatever activity in which humans, or any other beings with souls, engage themselves. Worst of all, the same research futility that plagued the physicist will return with a vengeance for the psychologist. Human psychology, if it can only recognize natural causes for events, will be forever on the hapless task of trying to explain the actions of the soul without including the soul in the theory.

The theistic evolutionist cannot go far enough to make his own position a persuasive one. He could banish divine action altogether and end up with a functionally useless God. That is unacceptable. The only alternative, however, is to allow God and the supernatural to intervene in the natural world at critical theological points. But once divine intervention is conceded in one place, why not at least look for it elsewhere? What harm can there be in having an open mind on the question? Theistic evolution is not impossible for a theist, just somewhat implausible.

Recent creationists have important religious and biblical considerations to buttress their suspicions of evolution, theistic or otherwise. They believe in a historical Fall with profound consequences for the entire natural economy. While they may differ on the physical character of the curse, recent creationists tend overwhelmingly to understand the curse of Genesis 3:14–19 as marking a radical change in the natural order, affecting not only human beings but all organisms. Death came to the world with Adam.

The secular account of natural history is a bloody one. It is full of dead-end species and waste. Evolution, in the modern sense, consumes life upon life to allow the fittest to survive and adapt. It is wasteful. It is inefficient. It is "red in tooth and claw."

With skillful argumentation this bloody history could perhaps be made consistent with the purposes of a wise and loving Creator. It is hard to see how such a story will ever be appealing to someone not already committed to Christian theism. The problem of evil in the world is hard enough to explain without the addition of millions of years of animal suffering to round it out. What is the justification for all that animal pain? After the Fall, this difficult problem can partly be laid at the feet of human beings. Our sin caused the pain of the world and its creatures. The naturalistic scientist, theistic or otherwise, has no such out. If death and extinction came before human sin, we cannot receive the blame for the millions of years of brutal suffering by countless beings! If the standard scientific chronology is true, then God willed it that way in an unfallen world.

This is not at all what one expects when one reads the biblical description of God. Again, God could have done things this way. His purposes are not always clear to us. But why believe such a thing of him? The biblical problems do not end here, however.

The Bible seems to teach that there was a global flood in the days of Noah. This was the universal teaching of the Fathers of the church. Though not directly linked to the issue of the age of the earth, one's position on the historical nature of the Flood and its extent are still important. The response one gives to this question will indicate important core religious ideas.

The sorts of issues that flow from the idea of a global flood are critical to a religious believer. What will control the biblical exegesis of the Christian? Will they forever be engaged in an exegesis of the moment? Later in this essay we will suggest an answer to these questions. For now, it is sufficient to make one simple point. Every Christian from the founding of the church until the advent of modern science believed Noah was a real person. The Catholic and Orthodox Churches venerate Noah as a saint with the other patriarchs.

Modern naturalistic science has found no room for a flood, global or local. Many Christians, even those otherwise quite conservative, suggest the Noah story is a myth. It contains important theological truth, but no history. The church was wrong. Noah never existed.

This is a serious move for the church to make. Do the considered opinions of scientists now have the last say in important religious matters that touch on history? To a secular person, Noah's disappearance looks very convenient. If a Bible story contains details that are contrary to science, then the Bible story is "myth." If the Bible story is fortunate enough to be unverifiable, like that of Abraham, it is allowed to function as history.

Henri Blocher and some other evangelical Bible scholars try to argue that this move away from history is justified by a proper reading of the text. Such a move is also perilous. It leaves the Bible a riddle to centuries of its readers who could not have understood it until now. In fact, so badly did God communicate that every reader thought it said something quite the opposite from what God meant. God did not mean for us to believe in Noah. He just wrote so badly, or in such an obscure idiom, that almost every other culture, in almost every place, thought Noah was a real man.

On the question of the earth's age, the evidence is much less clear. This has always been the case. The Bible no place states an age for the cosmos. This number can only be derived by extrapolation from the text. Such interpretation always opens up broader possibilities of error or misunderstanding of the text. Though most of the Fathers believed in a recent creation, they were not united in the way they understood divine revelation on this point.

Natural science at the moment seems to overwhelmingly point to an old cosmos. Though creationist scientists have suggested some evidences for a recent cosmos, none are widely accepted as true. It is safe to say that most recent creationists are motivated by religious concerns.

This need not be blind religious faith. It may be rational choice. Given the veracity of other seemingly implausible claims in Scripture, the recent creationist is willing to give other less vital doctrines the benefit of the doubt. To borrow language from philosopher W. V. Quine, we each have a web of beliefs to which we are committed. We hold to the central ideas firmly. Some of them are so firmly and rationally established in our thinking that to change our minds about them is (almost) unthinkable. Our other notions are based on these most important concepts. They may, in themselves, have little or no outside justification. They draw their strength from their association with the inner ring. We believe the story of Noah in Genesis draws most of its plausibility from the true story of Jesus in the Gospels.

The story of a man coming back from the dead is, after all, much less plausible than that of a global flood. Floods, even very big ones, do, after all, happen. A full-blown naturalist could, in principle, accept the idea of a *natural global flood*. But if naturalism is true, a truly dead man cannot live again. Any god capable of raising the dead is able to do a great deal. If Jesus can be shown to have risen, which we believe can be done, then many things that once looked impossible become interesting. The recent creationist has a right to see what can be made of the Flood story. Tracing such potential divine actions in the world and developing the theories to describe them will be hard scientific work. As we shall see when we turn to the philosophy of science, it is legitimate scientific work. The recent creationists may fail, but the effort itself will be intellectually interesting.

Why should we adopt recent creationism over an old earth model? We think that this question is much less important. At the moment, with both Christian and secular educational systems beset by naturalism, a truce is in order. The old earth creationist is an ally against both the theistic naturalism limiting the free flow of ideas inside the church and the secular naturalism cutting off new thinking in the universities. There are, however, two very good reasons to maintain a young earth position during the struggle.

First, recent creationism is intellectually interesting. Over the last twenty years, it has increased in sophistication and rigor. One needs to only read the papers from the last three meetings of the International Conference on Creationism to see the improvement. It has falsified some of its own ideas (e.g., the decline in the speed of light). It is listening to critics more and more frequently. Recent creationism even shows signs of becoming self-critical. Kurt Wise, the movement's leading paleontologist, is often brutally critical of recent creationist ideas. If it is a pseudoscience, it is the first one in history to evolve in this direction. It will be exciting to see where it ends up. In a postmodern world, we see no reason for traditional Christians to give up on an idea that intrigues them. As we shall argue later, if we are wrong, the universe will correct our errors sooner or later.

Second, a coherent recent creationism would be a great boon to religious belief. It is certainly possible to be an old earth creationist and be a good Christian. Recent creationism, however, is more interesting intellectually and religiously simply because it is more radical and ambitious in the scope of its projects. Imagine, if you dare, that Wise and his colleagues develop an empirically satisfying "flood geology." Go further and speculate about the impact of a successful recent earth cosmology explaining the data now viewed as indicating great age. There is no doubt that such a day would mark an important triumph for the faith.

Does this mean that we denigrate science and empirical research? Do we wish to "hold science hostage" to some outmoded reading of the Bible? Of course not. We do not wish free thinking held hostage by scientists or theologians. As we shall argue later, recent creationism is an attempt to reinterpret the data, not to deny their existence or importance. As it is now interpreted, the data are mostly against us. Well and good. We take this seriously. Eventually, failure to deal with that data in a recent creationist scientific theory would be sufficient reason to give up the project. We think, however, that progress is being made. The potential rewards outweigh the liabilities. Theistic naturalists and old earth creationists are free to develop the ideas. Recent creationists will do the same. In the end, we are confident that the world, and the Creator, will reveal the truth of the matter. In the meantime, dogmatic pronouncements from any camp are counterproductive. Recent creationists should humbly agree that their view is, at the moment, implausible on purely scientific grounds. They can make common cause with those who reject naturalism, like old earth creationists, to establish their most basic beliefs. When the dust has cleared from that intellectual revolution, they can then see how the landscape looks. It would not be surprising if many things once "known" for sure would be much less certain. This, of course, might include the age of the cosmos.

One other charge often hurled at recent creationism is worth a brief mention at this point. It is the *omphalos* (the Greek word for "navel") problem. Did Adam have a navel? This seemingly trivial question hides an important criticism often leveled at recent creationists.

If Adam was created with a navel, then he had the appearance of a history he did not actually have. He looked as if he had once had an umbilical cord and had been in the womb of a woman. However, being created from the hand of God, he had no such history. Thus Adam has an apparent history different from his actual one.

Recent creationists sometimes move from Adam's navel to cosmic history. Recent creationists have a problem with starlight and the size of the universe. It takes millions of years for light to travel from certain parts of our galaxy to earth. If the cosmos is only a few thousand years old, then such light could not yet have reached us. Yet such light is seen every night on earth (except for smog-laden L.A. and Chicago where we live). How does a recent creationist handle this problem?

Some suggest God could have created starlight in transit to the earth. Perhaps most of cosmic history is apparent rather than actual. Just as Adam had a navel and no mother, so starlight could exist without actually having traveled thousands of light-years. (Russell Humphreys and many other creationists are unhappy with this sort of argument. They are developing cosmological models that avoid the necessity of appealing to the appearance of history to explain things like distant starlight.)

Two charges are usually leveled at any use of appearance of history. First, some charge that this makes God a deceiver. Second, critics claim that it is improper to have most of cosmic history be apparent and not real.

The first criticism is not very compelling. One of our mothers loved to refinish and "antique" furniture. She would make a new chair look like an old chair. Was she deceiving her guests when she placed such a chair in the living room? Of course, first her guests would have to notice the chair. Then they would need to have the necessary knowledge about antiques to draw the wrong conclusion. If mother noticed the error and let it pass, she would be guilty of deception. She certainly would not be guilty, however, if she labeled the chair or pointed out the error to her guests. Recent creationists believe that the Bible and certain small indications in the natural world function as such a label.

The usual response to this defense brings us back to the second criticism. "Ah," says the critic knowingly, "we can see why your mother would antique a chair, but why would God create a universe that looked old?" The recent creationist could reply in two ways.

One way is to suggest that God needed such a creation to sustain life on earth. It might be necessary to have the universe the size and shape that it is in order for life on this planet to survive. Some strong form of the anthropic principle might demand a "big" universe. This "big universe" would need to be created in order to sustain life once the initial creative week was done. If God also wanted starlight to reach earth, he would have to create it en route. In other words, faraway stars, demanded by the need to have a universe, combined with a desire to a have a night sky full of "action," would lead to the creation of an apparent history.

The second approach to this problem would be to point out that God would have no real motive to "actualize" most of cosmic history. Before the creation of "free will" beings, cosmic history would (on a traditional Christian understanding of God's power and knowledge) be utterly foreordained. "Apparent" history in the mind of God could not be any different than "actual" history. There was, as yet, no sin or fallen man with which God needed to deal. For example, when watching a video, one frequently skips forward to the part one has not yet seen. In the same way, God might create the cosmos by skipping over the uninteresting parts (from his divine perspective). This would be prudent and efficient on his part. He would gain a fully functioning universe, but without the "waste of time" needed to actualize the less interesting parts. This speculation is merely to suggest that one could imagine good reasons for God to create with the appearance of history.

2. WHY IT MATTERS

Paul Nelson will begin to answer this question by describing a conversation that took place in Pittsburgh about twenty years ago. He had returned home on a holiday break from his first year of college. One evening he met with a handful of his closest high school friends to compare college adventures. After the usual roommate and professor anecdotes, the discussion wandered unexpectedly onto the topic of religion.

Paul was the only Christian in the room. When his turn came, he presented what seemed to him a strong case for Christianity, including vivid accounts of his spiritual experiences as a teenager, events as real to him as his own breath. He then sat back confidently to field the inevitable questions.

The first question stunned him. It came from "Melissa," a young woman with whom he had acted in school plays and shared many classes. She was attractive, brilliant, tall, and privileged. Sitting on the floor, her dancer's legs drawn up to her chest, she looked at Paul calmly and, without a trace of skepticism or insincerity, asked, "But why do I need to be saved?"

Paul cannot remember his answer. He cannot remember if he even attempted an answer. In the two decades since that evening, however, Melissa's question has often come back to him.

She was completely frank and completely mystified. Although Paul had described events of obvious significance to his own life, she could not understand why this made the least difference to her. Her response, note carefully, was not the bemused indifference of postmodernism, where it would not have mattered what Paul said, "truth" being a fiction in any event.

Rather, his story had quite simply been opaque from start to finish. Sin? Grace? Salvation? These words intersected with nothing Melissa recognized as real. The terms of his passionately rendered case for Christianity were empty. In the jargon of philosophers, they "failed to refer." What Paul had said might well have been interesting, even compelling in its own way, but examined in the light of Melissa's understanding of the world, none of it mattered.

What had stunned Paul should now be obvious. Suppose he had said, "I've just learned from the radio that a cloud of toxic gas is drifting toward us," or "I've discovered that the walls of the stage we shared for all those rehearsals exposed us to asbestos." Such statements would have conveyed, urgently, that Paul knew something that made a difference to everyone in the room. Melissa would have done something. She would have acted on what he told her.

"But why do I need to be saved?" however, did not move her to action. Compare Melissa's question to the one asked by the rich young ruler, the biblical character strikingly similar to her in gifts and social standing. The rich young ruler understood. "What good thing must I do to get eternal life?" (Matt. 19:16) he asked Jesus. Yet there is no reason to ask the question when one cannot see the point, or even grasp the meaning, of being saved at all. From what is one being saved? Why? What is Jesus talking about when he says, "Follow me"?

So why should anyone follow Jesus? There is a powerful picture of the universe, one held by most of the world's scientists and scholars, according to which this question—Melissa's question—makes perfect sense. One should follow Jesus only if what Jesus said was true. But now, in the fullness of our scientific knowledge, we know better. We know that the universe began as an uncaused quantum fluctuation. We know that galaxies, stars, and planets formed as the inevitable outcome of physical laws and chance events. We know that life first arose, whenever and wherever it did, as chemicals combined and recombined to cross a threshold of complexity into self-replication, and that every organism on earth is the naturally selected product of undirected variation among the offspring of those first self-replicating systems. And we know that somewhere in East Africa a few millions years ago, human beings came into existence by the same undirected process of variation and selection.

Of course, you will not hear an argument from the two of us that we truly know any of this. But suppose we did. Suppose this story really was true. Then, when Jesus said, "In the beginning, God created them male and female," to take an example close to home, it would appear that Jesus was mistaken. Or, if he knew better, that he was dissembling—"accommodating to his hearers" in more temperate language. Or, at the very least, that the relationship between Jesus' words and ordinary reality is a rather complicated business where one explanation, the scientific, satisfies the actual evidence at hand, and another explanation, the theological, refers to some other mode of being and explaining.

Melissa understood and accepted the scientific explanation. Her puzzlement stemmed from trying to discover exactly what meaning to attach to theology. Since Paul, who was trying to persuade her of Christianity, couldn't possibly have rejected the scientific explanation, and since he plainly did not think that Jesus was mistaken or was lying, he must have held the last view, namely, that the relationship between Jesus' words and reality was a rather complicated business.

And here, Melissa's common sense took over, as it has done for millions of people who find they lack the subtlety to follow out a complicated business. If we know that human beings evolved by an undirected process, then sin, for example, becomes an exceedingly difficult idea to get one's mind around. On this view, human behavior is, at bottom, the product of natural selection. We feel guilt because, as the Darwinian philosopher Michael Ruse puts it, our genes have tricked us. But, in the end, there is little more to say than that. In particular, adding "but God created us too" as a theological addendum to the naturalistic explanation invites the reply that one story is doing all the work, while the other is along for the ride.

But Paul might have said something else that evening. He might have said that we feel guilt when we do something wrong because God built us that way, by design, crafting and shaping us directly in his image. He might have argued that our moral sense could never have been generated by natural selection from any animal precursors, that no even remotely plausible evolutionary explanation exists for the origin of morality, and that whatever makes us moral creatures points to the handiwork of a Creator, before whom we are responsible.

Then, just as the Bible does, Paul would have staked a claim on Melissa's home turf—her reason. Maybe she would have been offended, and that would have ended the discussion. But he would have challenged, instantly, the wall of her intellectual self-confidence—in her case, the first step to reaching the core of her being.

Here is why this matters. The story of the Bible is embedded in history—in reality, the same reality, ultimately, to which natural science speaks. There really was a man named Jesus who died and was raised from the dead. But Jesus' life and words, in turn, are embedded in a larger story, which spills over at every point into what is commonly understood to be the province of science. The distinction between "matters of faith" and "matters of history and science" disappears when one grasps that faith takes as its object a man who spoke about history—the history of the people of Israel, the history of humankind, the history of the universe. When Jesus spoke of the Flood (Matt. 24:37–39) or of Adam and Eve, he endorsed the authority of the Old Testament, and his claims about himself only make sense in that light. One can imagine lines of authority running from Jesus out into the rest of the Bible, and, in turn, the rest of the Bible, beginning with Genesis, bearing on the life and ministry of Jesus as recorded in the Gospels.

3. PHILOSOPHY OF SCIENCE

Paul Nelson has a confession. Although he is trained in the philosophy of science, he doesn't much like the subject, or enormous tracts of it, at any rate. (A famous philosopher of science once told Paul that the last journal he ever cared to read was *Philosophy of Science,* the official journal of the Philosophy of Science Association, the discipline's biggest professional organization. "It puts me to sleep," he said.) This disaffection does not arise from thinking the philosophy of science a waste of time. Many scientists make that claim—the whole subject is a waste of time, they say—but they usually do so, oddly enough, in books or articles giving their own philosophy of science. "This subject is worthless," we might therefore take their condemnations to mean, "except when done by me."

No, the philosophy of science, in a sense we shall explain, is absolutely vital to learning the truth about the world. Here, then, is an easy way to grasp what nevertheless bothers us about the discipline, and why our response to this question takes the skeptically minimal form that it does.

Suppose you enjoy solving jigsaw puzzles, especially the most difficult ones with thousands of small pieces. As a useful rule of thumb, you adopt the practice of separating the edge pieces from the others and assembling them first. One day, as you start on a new puzzle, a couple of your friends stop to watch. "What are you doing with those pieces?" they ask. After you explain your customary practice of assembling the edge pieces first, they grow interested—not in the puzzle itself, however, but in the rule of thumb. "What is an edge piece, really?" one of them asks, holding a piece to the light. Soon they have forgotten entirely about the puzzle and are writing learned treatises on "The Conventionality of Jigsaw Piece Ontology" and "The Methodological Necessity of Initial Border Establishment in Puzzle Construction," with a lot of laborious squabbling about definitions and formal principles. ("An edge piece may be defined as such if and only if it possesses no less than one rectilinear dimension longer than each of its other dimensions.") Yet, when you venture modestly that one need not start with the edge pieces, since colors or patterns are also helpful sometimes, they whack you over the head with their learned treatises. "Surely you must know," they say, with all the offended dignity that academics can muster, "that the Edge Piece and Primary Assembly Principles are rationally indisputable in this community of investigators. What are you, some kind of crank?"

Of course, wondering what makes a piece an edge piece, rather than something else, might be what fascinates someone about jigsaw puzzles. Yet if that person thinks about or works on nothing else, he will have no time left to solve any actual puzzles. Let us say provisionally that science is a search for the truth about the natural world, which seems a reasonable thing to believe. We might then see that truth very much like a vast puzzle (call it "The Big Truth") made up of thousands of smaller particular truths carefully fitted together. If solving that vast puzzle, or some portion of it, is what really motivates you, then formulating and arguing about Edge Piece Definition and Primary Assembly Principles will be terribly distracting. Your attention may become consumed altogether by definitions and rule making—"contentions and barking disputations," in Francis Bacon's words—while the puzzle itself lies neglected.

But philosophical rule making of this sort can be more than a distraction. It may become an all-but-impassable obstacle to discovering the truth. Rule making may turn out to be an excuse for not thinking about issues that challenge the status quo. It can become a bad intellectual habit.

Bad intellectual habits, like bad habits in general, die hard. As the philosophers Larry Laudan and David Hull have argued, philosophical rule making in science (which, they note, has a long, if not entirely distinguished, history) tends to favor the rule maker. Indeed, creationists themselves have not been able to resist the temptation of trying to win disputes about matters of fact by the manipulation of verbal distinctions and gerrymandering of definitions. Many creationists have promoted philosophies of science which, not surprisingly, make evolutionary theory unscientific by definition. Notice: not *false* by definition. That would be a neat (but impossible) trick. *To know that something is false requires labor.* (Is Paul's e-mail address pnelson2@ix.netcom.com or pnelson3 @ix.netcom.com? We wish you the best of luck in using any definition to answer that question, which is incomparably easier to answer than most scientific puzzles. To find his correct e-mail address, you would need to leave your armchair and invest some effort.) Knowledge of truth and falsity in empirical matters, as we have been stressing, is gained only by work, not by verbal manipulations.

"But without some rules," worry the rule makers, "we will be faced with anarchy. Science must be governed by a common set of philosophical assumptions, or people will invoke whatever causes and principles suit their fancy, and the hard-won body of knowledge we call 'science' will decay into a chaos of unsupported opinion and prejudice."

Not at all. The world is a wonderfully stubborn place. No amount of trying will unscramble an egg. Arguments and debates among young earth creationists—over such questions as the decay of the speed of light or the existence of human footprints in the Cretaceous strata of the Paluxy River basin—demonstrate that even without the constraint of methodological naturalism, empirical inquiry will, as it were, *govern itself*. We need only trust that nature will talk back to us when we try to make her say something that isn't true. Nature herself, in a deep sense, provides all the constraints that science needs to generate new knowledge and lay aside what is false.

To see what we are getting at, consider another example from Paul's life. "A few years ago, my wife and I were preparing dinner for friends in an apartment in Cambridge, England, with a tiny, unused fireplace. Suddenly there was a loud commotion, and a pigeon, covered in soot, burst from the flue of the fireplace in a panic and careened around the room, banging into the ceiling until we managed to open a window and set it free. Some minutes later, our guests arrived. Without saying what had happened, we asked them to try to discover the cause of the fresh pattern of unusual marks on the ceiling."

Now suppose they had insisted, before looking at the evidence, that the cause must have been an artist's prank, or must have been a dirty tennis ball someone absentmindedly bounced off the ceiling, or a leak in the room upstairs, or any other particular sort of cause; and justified this restriction by a philosophical rule. That would dispatch the mystery, after a fashion—no one investigates something they think they already understand—but only at the intolerable cost of sacrificing the truth. Unlikely as it may seem, a pigeon really did fall down the chimney and fly out of the fireplace. Any rule that kept us from discovering that would only be an obstacle.

"Fair enough," say the philosophical rule makers. "No prior restrictions on scientific reasoning. We shall throw our thinking open to every possibility. Then how about the soot fairy? You know, a diminutive winged creature who covers herself with carbon debris and flies around inside apartments. Why couldn't we infer that as a cause?"

Well, how about the soot fairy? Having a philosophical rule such as "one cannot infer the action of a soot fairy" would make no difference to the facts themselves, if soot fairies actually existed. Nor would allowing for the possibility of soot fairies, if they did not. And indeed, there is no evidence for such creatures. But that is not something we could know, or learn, from a philosophical rule. We have to go out and look. We gain nothing toward discovering the true explanation—a pigeon fell down the chimney—by excluding possible causes before we investigate, even causes as silly and fantastic as soot fairies. In short, whatever philosophical rules we adopt before we investigate nature can only narrow what possibilities we may infer, but will never widen them. For there can be nothing more surprising, more beautiful and unexpected, than reality itself.

The reader needn't worry; we are not arguing that we look for fairies. Our point is simply that in excluding fairies, we might rule out a pigeon. The history of science is marked by many episodes where philosophical rules or a priori principles have kept scientists (or, as they would have called themselves before the mid-nineteenth century, natural philosophers) from discovering new truths. Galileo, for instance, advocated the Copernican theory, which held that the sun, not the earth, stood at the center of the universe. This was a sharp break from the then-dominant Ptolemaic-Aristotelian cosmology, in which the earth was motionless while the heavens rotated around it. But both Copernicus and Galileo, rather more than they knew, were also students of Aristotle. It was a principle of the Aristotelian world-system that heavenly bodies moved with only circular uniform motion. "The mind shudders," wrote Copernicus in his masterwork *De Revolutionibus* (1543 [1990, 514]), at any theory postulating noncircular, nonuniform motion, "since it is quite unfitting to suppose that such a state of affairs exists among things which are established in the best system." Galileo himself wrote, in his *Dialogue Concerning the Two Chief World Systems* (1632), "I therefore conclude that only circular motion can naturally suit bodies which are integral parts of the universe constituted in the best arrangement."[1]

[1]Galileo Galilei, *Dialogue Concerning the Two Chief World Systems* (Berkeley: University of California Press, 1967), 32.

Circles it must therefore be, for circles it can only be—except, of course, that the planets actually move in ellipses, accelerating as they approach the sun. To discover this, Johannes Kepler, a contemporary of Galileo, had to ignore or reject the very philosophical principle that Copernicus and Galileo held to be rationally indisputable. "Circles should not be used," Kepler argued, "because they were made for laying physical explanations of movement before the imagination rather than astronomical [evidence]." Kepler ignored the rules to get to the truth.

This kind of philosophical rule making has been the occasion for truly outrageous behavior in the academy. Here, for instance, is some advice about the creation-evolution controversy that a major scientific organization, the American Association of Physical Anthropologists (AAPA), actually offered its members:

> In any confrontation, you should be prepared to show that evolution is scientific, not that it is correct. . . . One need not discuss fossils, intermediate forms, or probabilities of mutation. These are incidental. The question is, What is science and what is religion. Therefore, if you must confront the creationists, we suggest you discuss the nature of science, the kind of knowledge it can provide, and the kind it cannot provide.[2]

Notice what has happened. Anyone who asks about the truth of evolution will, as a matter of philosophical definition, be shunted onto a dead-end siding. There that person will be invited to contemplate, not the physical evidence—it being "incidental" in any case—but "the nature of science." This is the purest sophistry.

Religious people can be tempted into such questionable maneuvers as well. If the Christian researcher comes to the field sure of what God would or would not do, then he or she may miss what God really did. This affliction can strike the young earth creationist, tempted to shoehorn contrary data into her "flood account." Facts that should induce further research or a new paradigm may be ignored as inconvenient or irrelevant.

---

[2] "A Recommendation to the Association Concerning Creation," *American Journal of Physical Anthropology* 62 (1983), 457–58.

Argument by definition can similarly inhibit the science of a theistic evolutionist. The theistic evolutionist knows that God would not "intervene" in the natural world. He knows that God would create matter with certain "capacities." Such a person may be unable to see the positive evidence for design in the natural world. He is blinded by his rule making from doing good science.

4. THEOLOGY AND SCRIPTURE

We believe that Scripture and church tradition provide us with good reason to believe at least five important things about humans and the cosmos. As Nancy Pearcey and Charles Thaxton pointed out in *The Soul of Science*,[3] these ideas contributed to the birth of modern science in the first place. These ideas are:

1. *The world is created by an intelligence or mind in whose image we are created. Therefore, it is contingent.* Both facts together make us hopeful as humans that our minds can empirically examine and know the world, even if imperfectly.
2. *Our knowledge is incomplete.* Even the seemingly best-supported human idea may be wrong.
3. *God is absolutely free.* This leads to an openness on our part to all possible modes of causation. God could allow the universe to function based on its "creaturely capacities" or he could actively intervene. This means that design is an empirical possibility.
4. *The Bible is true.* If it describes an event and asserts it happened, then it happened. It seems very implausible to us that God would have used numerous false stories to convey his message. Scripture does not read like the *Timaeus* or *Epic of Gilgamesh* or other ancient stories.
5. *There is a moral dimension to all knowing: science is not metaphysically neutral.* There is such a thing as intellectual sin.

Here we might sound a note of realism about philosophical naturalism. Because God created an ordered universe marked by regularities, scientific inquiry can indeed go quite far, successfully, by concentrating its attention on "natural laws," mechanisms, and regularities. But it is easy to elevate these

---

[3]Nancy R. Pearcey and Charles B. Thaxton, *The Soul of Science: Christian Faith and Natural Philosophy* (Wheaton, Ill.: Crossway, 1994).

things to an autonomous status, that is, to turn God's created order into a brute given, to make it the ultimate reality. Indeed, it is possible, to use the blunt and unmistakable language of the prophets and the apostle Paul, to make an idol of nature.

One can conceive this with another metaphor. Recall that we said science ought to seek the truth about the natural world. Now there are many truths fully consistent with both methodological naturalism and theism (e.g., that blood circulates in vertebrates, that the moon orbits the earth, and so on). But there are other truths that naturalism denies, which, in fact, it is incapable of expressing. And here we find a deeply significant asymmetry between naturalism and theism, which deserves a name. We propose one as follows.

In Christopher Marlowe's *The Tragical History of Doctor Faustus,* Professor Faustus strikes a bargain with Lucifer, who sends him the devil Mephistopheles as a servant. Reveling in his power over Mephistopheles, Faustus puts questions to him about the cause of eclipses and other natural phenomena. ("Why have we not conjunctions, oppositions, aspects, eclipses, all at one time, but in some years we have more, in some less?") Then Faustus grows more bold:

> Faustus: Tell me who made the world.
> Mephistopheles: I will not.
> Faustus: Sweet Mephistopheles, tell me.
> Mephistopheles: Move me not, for I will not tell thee.
> Faustus: Villain, have I not bound thee to tell me anything?
> Mephistopheles: Ay, that is not against our kingdom, but this is.

On first reading this passage years ago, both of us thought Mephistopheles had contradicted himself. How could he agree to tell Faustus "anything," and yet refuse to answer his question about the authorship of the world? But Mephistopheles has not contradicted himself. Rather, there is an ambiguity in the word "that" ("that is not against our kingdom"). Of course, the "anything" that Mephistopheles has promised to tell is not truly anything, but only that which is not against "our kingdom"—that is, the kingdom of Lucifer.

If the world was created by God, however (which Faustus, who for the moment is under no similar constraint himself, exclaims in his next line: "Think, Faustus, upon God that made the world"), then the knowledge promised by Mephistopheles is a shabby, small, desperately cheap and inconsequential portion of an incomprehensibly greater kingdom of truths. Notice the wretched deal Faustus has made for himself. The questions he puts to Mephistopheles about eclipses and the like would still have the same answers had he never made a pact with Lucifer; that is, the answers would be contained in the wider kingdom of truths about which Mephistopheles must be silent. It would seem that Mephistopheles can tell Faustus nearly everything. In fact, if God created the world, he can tell him almost nothing.

Let us call this the Mephistophelian Asymmetry. The asymmetry obtains between a theism that can recognize and use any natural mechanism or regularity for which there is good evidence, *as well as the hypothesis of intelligent design*, versus a methodological atheism (or naturalism) which is, by its own rules, restricted solely to natural laws and chance. To stand on the theistic side of the asymmetry places one in a far richer sphere of possible causes.

The asymmetry obtains between the truths there are to be known, and the truths we have allowed. Something very like the Mephistophelian Asymmetry exists with philosophical naturalism in the following sense. Now, the kingdom of philosophical naturalism, to be sure, does not have Lucifer at its head. Rather, it takes the physical, material universe as comprising everything that there is.

There are scientists, including some who are Christians, who think that such a bargain is necessary. One needs to make the deal in order to avoid the terrors of a "God of the gaps." The God-of-the-gaps fallacy has become so terrifying a monster that, at the mere mention of its name, otherwise stalwart scientists and philosophers scurry for the safety and cover of such philosophical doctrines as methodological naturalism. "You won't catch *us* committing a God-of-the-gaps blunder!" they call from their coverts. "We see only natural laws and regularities around us. And if we don't—well, those laws or regularities will turn up, eventually. Whatever the appearances, we must never assume that there are gaps in the natural order into which a God might be inserted. That's the dread fallacy!" Yes, yes—the dread fallacy! One hears the echoing shouts from other hiding places in the open field.

It may be impossible to coax from their philosophical coverts those who have been forever intimidated by the monster. However, it is possible to lead the God of the gaps by the hand, so to speak, to a place of prominence in the field, where the monster, who is actually rather tame and mild, can be dismantled piece by piece.

The first piece, and the easiest to remove, is the term "gap." To what does this refer? Here we should consider a vital distinction. On the one hand, we have the world and its phenomena (i.e., the data we wish to explain). On the other hand, we have our theories about the phenomena (e.g., the Big Bang, neo-Darwinism, various accounts of creation). The "gap" in the phrase "God of the gaps" does not refer primarily to the phenomena, although it does bear some relation. Rather it refers to a gap in our *understanding,* namely, in what our theories *about* the phenomena have led us to expect.

Consider a nonscientific example. Suppose you visit a store in a distant neighborhood on your way home from work. It's your first time in this particular store, and you intend to buy your usual load of groceries. Only this store stocks just a few apples in its produce section, carries no canned food, and has a vast array of baked goods. So you complain to the manager, "Hey, where's the lettuce and the canned soup? There are serious *gaps* in this supermarket! All this bread, but no lettuce or soup?"

"Sorry," says the manager with a shrug, "but the gap is in your head. Every now and then, we have a bit of fruit for sale. But this is a bakery."

Since the customer assumed that the business was a supermarket, the customer expected it to carry certain items in its inventory. For sake of example, call this the customer's "theoretical expectation." Remove that particular expectation, however, and you will eliminate the perceived gaps. In other words, the gap about which the customer complained was unreal because it existed not in the world but in the customer's own understanding. Thus the customer had no reason to expect to find certain groceries since a bakery carries baked goods, not lettuce and canned soup.

In science, as in life in general, theories create expectations. They create gaps; but the gaps exist primarily in our understanding about the world. The gap is given by, or is relative to, some theory *about* the phenomena. When a scientist worries that someone may be committing the God-of-the-gaps fallacy, it is because a scientific theory he accepts has told him to expect to find an answer—typically a natural regularity or mechanism—where someone else has inserted the agency of an intelligent designer, or God.

But that scientific theory may be wrong. If it is, the gap in his understanding, which he hopes to fill with an as-yet-undiscovered law or mechanism, may be illusory as well. He may be looking for lettuce in a bakery.

If you consider the contrast between what the world is really like and what our potentially flawed theories tell us it must be like, you will save yourself from a pointless intellectual struggle. Consider, for instance, the problem of the naturalistic origin of life. This gap arises for evolution because, according to that theory, the parts must precede the whole, that is, the non-living constituents of organisms must be temporally and causally prior to organisms. In other words, first came methane, carbon dioxide, ammonia, and water; from these came amino acids; from these came proteins, and so on. In broad outline, that is the theoretical story. Thus the task facing researchers is to fill in the mechanistic gaps without resorting to a designer, a cause forbidden by the rule of methodological naturalism.

But nothing in the question "How did organisms first come to be?" dictates that the parts of organisms must precede the whole. Indeed, quite the opposite may be the case. With biological systems the whole may precede and causally govern its parts and, in the absence of some sound philosophical justification, the prohibition against intelligent causation is arbitrary and question-begging.

If we take a design-based theory as our guiding picture, however, the gap created by the evolutionary puzzle of how life arose through natural mechanisms wouldn't be so much filled by design as dissolved by it. As the philosopher Thomas Kuhn writes of problems posed by theories in general, the question of how life arose naturalistically is a puzzle "for whose very existence the validity of the paradigm must be assumed. Failure to achieve a solution discredits only the scientist and not the theory. Here . . . the proverb applies: 'It is a poor carpenter who blames his tools.'" It is not a poor carpenter who blames the blueprint, however, when it dictates an impossible structure. It is rational for a scientist who recognizes the intractability of his research puzzle to abandon it, if he discovers that the puzzle presupposes something false. (It would be irrational to do otherwise.) Likewise it is rational for a scientist to cease to try to fill a gap that does not exist.

The gaps in our understanding depend on what theories we accept—and the theories we accept will depend critically on what assumptions we use. Consider, for instance, a key assumption of standard cosmology:

> The grand hypothesis that nearly every cosmologist makes is that the universe (on a grand cosmological scale larger than 10^8 light-years) looks the same no matter where one may be in it—although not necessarily simultaneously. That is, at any point in the universe there will be a time, or there was a time, at which the universe looked or will look as the universe looks to us now. . . .
>
> The assumption we have just mentioned implies a very strong uniformity in the universe. It is a completely arbitrary hypothesis, as far as I understand it—and of course not at all subject to any kind of observational checking, since we have been and will continue to be confined to a very small region about our galaxy, and the time development of the universe follows a "cosmological" scale a billion times longer than our lifetime. I suspect that the assumption of uniformity of the universe reflects a prejudice born of a sequence of overthrows of geocentric ideas. When men admitted the earth was not the center of the universe, they clung for a while to a heliocentric universe, only to find that the sun was an ordinary star much like any other star, occupying an ordinary (not central!) place within a galaxy which is not an extraordinary galaxy but one just like many many others. Thus, it is assumed that our place in the universe should be just like any other place in the universe, as an extension of the sequence I have described. It would be embarrassing to find, after stating that we live in an ordinary planet about an ordinary star in an ordinary galaxy, that our place in the

> universe is extraordinary, either being the center or the place of smallest density or the place of greatest density, and so forth. To avoid this embarrassment we cling to the hypothesis of uniformity.
>
> Yet we must not accept such a hypothesis without recognizing it for what it is. An analogy will illustrate my viewpoint. If we parachute out from a plane flying at random over the earth and land in a clump of birch trees about this spot we might argue, we landed at random in no particular spot—there being nothing unique about this spot we conclude that the earth is covered with birch trees *everywhere!* The conclusion would be false regardless of the perfect randomness of the place where we might land. Yet perhaps we are doing the same thing in constructing our fundamental assumption of cosmology.[3]

The point is simply that the recent creationist should be allowed to challenge some of those basic tenets of cosmology, if doing so will enable him to produce a coherent and satisfying worldview embracing both religious and scientific points of view.

5. EPISTEMOLOGY[4]

We can best answer the question of the apparent tension between science and theology by first showing one method that should not be used to resolve Bible and science conflicts. In his work *The Authority of Scripture,* Galileo Galilei wrote,

> Since the Holy Writ is true, and all truth agrees with truth, the truth of Holy Writ cannot be contrary to the truth obtained by reason and experiment. This being true, it is the business of the judicious expositor to find the true meaning of scriptural passages which must accord with the conclusions of observation and experiment, and care

---

[3]Richard P. Feynman, Fernando B. Morinigo, and William G. Wagner, *Feynman Lectures on Gravitation*, ed. Brian Hatfield (Reading, Mass.: Addison-Wesley, 1995), 166–67.

[4]This section of this essay reflects an earlier version that appeared as the conclusion to a much longer paper. See John Mark Reynolds, "The Bible and Science: Toward a Rational Harmonization," in the *1994 Proceedings of the International Conference on Creationism* (Pittsburgh, Pa.).

must be taken that the work of exposition does not fall in to foolish and ignorant hands.

Galileo then proceeded to harmonize the account of the sun standing still found in the book of Joshua with his heliocentric view of the cosmos.

This method of harmonizing Scripture with empirical data makes the interpretation of the Bible dependent on scientific fact. Roughly speaking, Galileo would have the Christian read the Bible through the lens of empirical data. This has become a common approach to the interpretation of the Bible within certain Christian circles.

This view has one basic advantage. It gives the person approaching the Bible a means to solve certain interpretative puzzles. The legendary arguments between a flat or spherical earth are a case in point. Either reading could find scriptural support. For example, the Bible speaks of the "four corners of the earth" in Revelation 7:1. This could be interpreted poetically, as all evangelical commentators now do, or it could be interpreted literally. If interpreted literally, one could extrapolate from this reference to a rectangular earth. On the other hand, Isaiah 40:22 says, "He sits enthroned above the circle of the earth . . . ," which some interpreters may view as textual support for a spherical earth. This is especially true if it is read in the context of Job 26:7: "He spreads out the northern skies over empty space; he suspends the earth over nothing."

How does one decide which way to read the text? For Galileo, the solution was simple. Modern science has determined that the earth is spherical. Since Scripture is true, it must *actually* teach a spherical earth. All references that seem to imply otherwise must be harmonized with the scientific data. Where science is silent, the exegete is free to pick the most natural interpretation of the text.

This methodology, however, has two essential problems. First, it makes Scripture potentially nonfalsifiable. Second, it frequently fails to take into account a distinction between *observations* and the *conclusions based on observations*. It is overly simplistic and too trusting of whatever scientists happen to accept.

There is something troubling about the fact that there is no built-in limit to the amount of accommodation possible. What is meant by the statement "The Bible is true" if accommodation proceeds past a certain point? People holding this view, or any view like it, would need to clarify how far they are willing to stretch language before giving up the initial premise. As the argument stands now, the Bible could theoretically be made to say the opposite of its "plain sense" and still be defended as "scientifically accurate." This is disconcerting.

Of course, the same argument applies to people who view as myth those parts of the Bible that are "refuted" by modern science. Many theistic evolutionists, to their credit, believe in a literal, physical Christ leaving the tomb. Conveniently, this is a miracle unlikely to be within the reach of modern scientific examination. What can we make of a belief system that accepts some parts of salvation history as literal (the parts we can't empirically examine) and makes some parts mythical (the kind we can examine)? It is hard to see any great evangelistic force to it. It might save one frightened generation of Christians, but it is unlikely to bring many new people to the faith.

Galileo and his kind argue:

1. The Bible is true when it describes the world.
2. Certain facts of science are true about the world.
3. In every case, if two things are both true about the world, then they do not contradict each other. Therefore,
4. The facts of science and the descriptions of the Bible do not contradict each other.
5. When the facts of science and the descriptions of the Bible do seem to contradict each other, then the descriptions of Scripture must be reinterpreted in light of point 4.

There is a twofold problem with this argument. First, it depends on the unstated philosophical assumption that the empirical data that reach the senses are more reliable than the human interpretation of Scripture. This is a debatable presupposition. We can grant it, however, for the sake of argument.

The second problem with the argument is the failure to distinguish between an observation of science and a conclusion or theory of science based on the observation. The argument assumes that all scientific disagreement must be decided by changing the interpretation of Scripture. But is this plausible? Perhaps some theories of science are tenuous enough to make one wonder whether an elegant biblical exegesis should be abandoned when they conflict. Certain facts of experience (e.g., objects generally fall when dropped; the world is roughly spherical) seem to be more intuitively certain than certain other human interpretations of data (e.g., the cosmological assumptions noted earlier).

Galileo appears to have forgotten that just as Scripture is dependent on a hermeneutic for understanding, so science is dependent on an interpretive framework for comprehensibility. All the "facts" or "theories" of science do not have the same epistemic certainty.

A form of Ockham's razor would lead us to prefer simple scientific theories to complex ones. We do not want to needlessly complicate our science with useless entities. The same rule, however, should also be applied to the reading of Scripture. The simplest exegesis of Scripture is also to be preferred. By simplest, we mean the most natural or the one with the most external linguistic support. Perhaps in balancing the two, a slightly less elegant scientific theory that preserves a remarkably elegant biblical reading is preferable to a slightly more elegant scientific theory that produces a tortured exegesis.

For Galileo, the truth of Scripture and the phenomena of science carry equal weight. What has been assumed is that the "facts of science" are just that—facts. In most cases, the facts of science are interpretations, which are more or less plausible for explaining certain phenomena. The phenomena themselves have a high degree of intuitive plausibility, but the theories have much less. As W. V. Quine suggests, it is perhaps possible that two mutually contradictory theories could be postulated to explain the phenomena.

Theories of science are epistemologically similar to interpretations of Scripture. Both are good insofar as they explain the data at hand. It is *not* always a matter of some brute fact conflicting with a particular reading of Scripture but one interpretative framework confronting another.

We would like to suggest that scientific "fact" and "theory" operate more on a continuum. There are certain brute facts about the world. These are very certain descriptions about the cosmos. An example might be, "there is a tree." These should take priority over any interpretation proposed of Scripture. If the Bible seems to say that there is not a tree over there, then (if the Bible is true) the simplest exegesis would have to be modified. An actual biblical example might be Jesus' description of the mustard seed. He described this seed as the "smallest" of all seeds (Matt. 13:32). In fact, the most natural interpretation would be that there were no known seeds smaller than the mustard seed. Of course, science might find a smaller seed. In this case, the "fact of the matter" and an interpretation of Scripture conflict. Scripture must be reinterpreted if it is to remain "true."

On the other hand, some natural interpretation of Scripture based on the whole of the text might conflict with some theory of science. *If there is another interpretation of the data that preserves the natural meaning of the text, it is to be preferred.*

These ideas could be expressed in the following model of interpretation that is based on two scales:

A. Biblical Knowledge	*B. Scientific Knowledge*
III. Allegorical/Mythological (possible)	III. Highly Complex Theory (possible)
II. More Complex Exegesis (plausible)	II. Fairly Complex Theory (plausible)
I. Natural Exegesis (probable)	I. Simplest Theory (probable)
Raw Data of the Text	*Phenomena of the Cosmos*

The scale would allow for gradations between points. The object of the person harmonizing Scripture with science would be to get the simplest scientific theory in combination with the most natural possible exegesis.

If at all possible, the extreme end of the scales should be avoided (viz., Allegorical/Mythological interpretations and Highly Complex Theories). A theory that contains a major component that is barely possible (despite the fact that the other major component is a probable explanation) is not as intuitively satisfying as a theory that contains two fairly plausible components. Hence, on the assumption that both natural and divine revelation are equally valid, one would desire as balanced a view as possible.

We do not claim that this would resolve all conflicts between science and Scripture. The question of whether a probable reading of Scripture combined with a plausible scientific theory would be preferable to a probable scientific theory combined with a plausible reading of Scripture is not yet resolved. For example, we would take the argument over the day-age interpretation of the Hebrew word *yom* in Genesis 1 to be such an issue.

It would resolve many types of conflicts between science and Scripture. A case in point is the historic controversy over the movement of the earth. The majority of the evidence of the Bible seems to support a still earth on the most natural reading of the text. Such a reading would be the result of natural exegesis, a "probable" reading. For some time the Aristotelian theory of the cosmos was a "probable" scientific theory. No conflict existed between science and Scripture: both divine revelation and scientific theories were in the "best" positions. Following the introduction of evidence supplied by certain late Medieval natural philosophers, the Aristotelian theory moved from being the most probable theory to being a plausible one. However, we believe the church was right in maintaining the classic "unmoved earth" position at that stage of the dialogue. Only when such a position became mathematically and observationally "hopeless," should the church have abandoned it. This, in fact, is what the church did.

Young earth creationism, therefore, need not embrace a dogmatic or static biblical hermeneutic. It must be willing to change and admit error. Presently, we can admit that as recent creationists we are defending a very natural biblical account, at the cost of abandoning a very plausible scientific picture of an "old" cosmos. But over the long term, this is not a tenable position. In our opinion, old earth creationism combines a less natural textual reading with a much more plausible scientific vision. They have many fewer "problems of science." At the moment, this would seem the more rational position to adopt.

Recent creationism must develop better scientific accounts if it is to remain viable against old earth creationism. On the other hand, the reading of Scripture (e.g., a real Flood, meaningful genealogies, an actual dividing of languages) is so natural that it seems worth saving. Since we believe recent creation cosmologies are improving, we are encouraged to continue the effort.

A LETTER TO SUSAN

Dear Susan,

You face some difficult decisions. Take your time. Don't jump into a "camp" too quickly. Allow yourself the time to read the major books for each position. Realize that each group has its strengths and weaknesses. Weigh which arguments are most persuasive to you.

Avoid two temptations right from the start. (Our warnings come from personal experience!) First, avoid the snare of "the safe." Don't pick a view just because it will make your parents and your church happy. Also, avoid settling on a set of ideas that will "get me though graduate school." What is the profit in getting a diploma if you have to turn off your brain to do it?

Don't assume that a position is correct merely because most practitioners of science believe in it. Remember that critical thinking skills and logical analysis are not the domain of any one discipline. Every discipline's arguments must be subjected to these rules. When Richard Dawkins makes a bad argument using science examples, it is still a bad argument despite his skill in science.

One of the authors of this essay (John Mark Reynolds) went to college as a theistic evolutionist. He had not been raised to see any particular tension between religion and science. When his comfortable views were subjected to the cross fire both of Christian and secular friends, he discovered that theistic evolution often consists of weighty adjectives and theological jargon marching no place in particular. What does it mean for God to sustain the universe? How plausible is evolution as a creative mechanism for God? Why should we see "creaturely capacities" in matter instead of simple "capacities"? To his dismay, John Mark discovered that in theistic evolution, the theism did not do very much metaphysical or physical work outside of the head of the believer.

Theistic evolutionists often pretend that their ideas are not new. Go read the church fathers for yourself. Read Augustine and Basil. Read John Chrysostom and the Eastern Fathers. See what each had to say about the Flood and about Creation. Ask yourself if theistic evolution fits the great tradition of the Christian faith. It does not. It would be a shame to abandon that tradition without even knowing it or without great cause.

Most importantly, however, think about whether evolution *really happened*. Sure, it could have happened. God could do anything. That is not the point. What is the status of the evidence? We believe that a fair reading of the evidence, one that does not put on philosophic or religious blinders, will give you good reason to doubt the truth of Darwinism.

That is not enough, however. You will quickly see you don't have to be a Darwinist to be rational. What should you be? Take your time with that as well. Just as Darwinism was a powerful modification of some very old ideas (going back as early as the fifth century B.C.), so new design arguments are being developed. They are not like those of Plato or Paley. They avoid the old errors while keeping the strength of the notion of intelligent design.

We believe the intelligent design movement is on the cutting edge of what is happening in religion and in science. Feel free to put off the question of the age of the earth and the Flood for now. Young and old earth creationists agree on far more than they disagree. Avoid people in both camps who call names. Be open to the idea that when dust settles from the fall of Darwinism, young earth creationism may have some appeal to you. Keep an open mind.

Socrates constantly encouraged the young people around him to engage in the life of the mind. "You inquire!" he would say to his disciples. Adopt that as your motto. Don't let a religious or scientific establishment stamp you with an unthinking orthodoxy. Be properly skeptical.

We both have found that when you are willing to take intellectual risks, the reward is very great. Being part of the intelligent design movement as well as being open to even more radical possibilities is liberating. Like Plato, we believe that such a pursuit of truth ends in a vision of the good, the true, and the beautiful. The sophists in the academic establishment will thunder against you. A glimmer of the truth is worth the price. Why? Because the truth is not just a good idea—the truth is a Man. And when we know him and abide in his Word, then we will know the truth and he shall set us free.

RESPONSE TO PAUL NELSON AND JOHN MARK REYNOLDS

Walter L. Bradley

I will concentrate my critique of Nelson and Reynolds on the scientific issues raised in their chapter, leaving the philosophical and theological critiques to my colleagues. Nelson and Reynolds summarize the distinctive points of young earth creationism as (1) an open philosophy of science; (2) the direct creation of basic animal types as described in Genesis 1 and 2; (3) the curse of Genesis 3:14–19 and its profound effect on every aspect of the natural economy, but especially the introduction of physical as well as spiritual death at the time of the Fall; and (4) the historical and worldwide flood of Noah. Implicit in these distinctives would be a recent creation with the large fossil record attributed to the Flood.

Points 1 and 2 would be common to young earth and progressive creationists, points 3 and the requirement of a recent creation it implies would be differences between these two groups, with all three points being differences between young earth creationists and theistic evolutionists (also referred to in this book as the "fully gifted creation" position). A historical flood would probably be a point of agreement between young earth and progressive creationists and theistic evolutionists, but the extent of the Flood would be the point of contention. I agree with the various comments by Reynolds and Nelson in support of points 1 and 2, and thus will not be redundant in discussing these points further.

The question of what changes in the natural world took place at the Fall (their point 3) is a difficult one. If death was introduced for the first time after the Fall, then all of the apparent evidence of death in the fossil record must be attributed to the Flood. However, Dan Wonderly[1] and others have clearly documented that fossils in sedimentary rock in the Gulf of Mexico go down 25,000 feet, with many different types of sedimentary rock (e.g., from shells, from deposits of soil, from precipitates, from evaporation of shallow seas), each formed under very different geological conditions and each with ample fossils. It is difficult to account for this stupendous historical zoo from one recent, worldwide flood. This situation is repeated around the world, making it extremely difficult to accept the fossil record as the consequence of one worldwide flood. It is also worth noting that some necessary biological activity in human beings, such as bacteria in our stomachs, would seem to be impossible if no death of living organisms before the Fall is permitted.

In my opinion, the interpretation of Genesis 2–3 as referring to a physical fall is itself problematic. The text in Genesis 2 indicates that in the day that Adam or Eve would eat of the fruit of the tree they would "surely die" (v. 17 KJV). Yet physical death came much later to Adam and Eve; only spiritual death came at the time of the Fall. This would seem to imply that the death alluded to in Genesis 2 was spiritual, not physical. Furthermore, since this world has always been temporal, it is problematic that Adam and Eve would have lived "forever," even if they had not eaten of the fruit of the tree.

Sometimes young earth creationists (but not Nelson and Reynolds) suggest that the second law of thermodynamics was the curse put on nature at the Fall. However, this hypothesis fails to recognize the many necessary physical processes, especially transport phenomena, that depend on the second law of thermodynamics. In the absence of the second law of thermodynamics, we would suffocate in the carbon dioxide we exhale since it would not diffuse away. The curse on nature at the time of the Fall described in the Bible is real, but it must be something other than physical death or the second law of thermodynamics.

---

[1]Dan Wonderly, *God's Time-Records in Ancient Sediments* (Flint, Mich.: Crystal Press, 1977).

Finally, the young earth creationists like Reynolds and Nelson, with their model of a recent creation and no death before the Fall, must attribute most of the fossil record around the world to the Flood, making this a scientific necessity in addition to a biblical necessity in their view. However, the intent of the Flood in Genesis would seem to be accomplished by the extent of the Flood being as widespread as Adam's descendants, which might not have been worldwide at the time of the Flood. Scientific evidence for a regional flood is quite strong, while evidence for a worldwide flood is less evident. I am ambivalent on which it is.

The other major problem with young earth creationism is the overwhelming scientific evidence for an old earth. This has been well documented in several books in recent times.[2] The size of the universe, the many different radiometric clocks, the development of coral reefs, and thousands of feet of widely varying sediment all argue convincingly for an earth and a universe of great antiquity. To their credit, Reynolds and Nelson acknowledge this current state of affairs regarding the scientific evidence, without trotting out some of the usual, discredited scientific arguments for a recent earth.

The events of Genesis 2 which recap a portion of the sixth "day" do not neatly fit into a morning-to-evening scenario. Thus, it is questionable whether the young earth interpretation of Genesis 1–2 is preferable. Even the International Council for Biblical Inerrancy in their 1982 conference on hermeneutics refused to affirm the necessity of interpreting Genesis 1 as a recent, six-solar-day event.[3] In an invited paper for this meeting, I weighed the scientific and biblical evidence and concluded that a young earth position was less plausible than a progressive creationist position.

Reynolds and Nelson conclude their discussion by offering the reader a choice between a probable reading of Scripture and

---

[2]Davis A. Young, *Creation and the Flood* (Grand Rapids: Baker, 1977); Hugh Ross, *Creation and Time: A Biblical and Scientific Perspective on the Creation-Date Controversy* (Colorado Springs: NavPress, 1994).

[3]Walter L. Bradley and Roger Olsen, "The Trustworthiness of Scripture in Areas Relating to Natural Science," and "Chicago Statement on Biblical Hermeneutics," in *Hermeneutics, Inerrancy and the Bible*, ed. Earl D. Radmacher and Robert D. Preus (Grand Rapids: Zondervan, 1984).

a more plausible scientific theory (their view of progressive creationism) and a plausible reading of Scripture with a probable scientific theory (young earth creationism). I would argue that their claim that a young earth creationism is a more plausible interpretation of Scripture is dubious.[4] Also, I would add that the scientific plausibility of progressive creationism is overwhelmingly superior when compared to young earth creationism. Later I will summarize my preferences for progressive creationism over theistic evolution in my critique of Howard J. Van Till's chapter.

[4] Bradley and Olsen, "The Trustworthiness of Scripture."

RESPONSE TO PAUL NELSON AND JOHN MARK REYNOLDS

John Jefferson Davis

I find myself in agreement with a number of significant points made by Nelson and Reynolds in their chapter on "Young Earth Creationism." I understand their reservations about this terminology, but continue to use it as they themselves do, given the wide currency of this usage in the literature.

I certainly would affirm what they designate as an "open philosophy of science," that is, a philosophy of science that does not exclude the supernatural and that recognizes the ability of the Creator to intervene directly in the natural order. Neither I nor these authors would support a "methodological naturalism" in the philosophy of science that would attempt to define as illegitimate any references to God as possible explanations for observed features of the natural world. I would affirm the point made by Alvin Plantinga in this regard, namely, our philosophy of science should incorporate all that we believe to be true about the world, including the allowance of divine revelation as well as reason and the scientific method as possible sources of information.[1]

The authors and I also share a common affirmation of an intelligent design approach to many of the questions surrounding issues such as the origins of life on earth and explanations of the complexity of living systems. We share a common understanding that a naturalistic principle of natural selection is in

[1]Alvin Plantinga, "Methodological Naturalism?" *Perspectives on Science and Christian Faith* 49 (September 1997): 143–54.

itself inadequate to explain the origins of the first living cell and the irreducible complexity observable in the biological world.[2]

I also salute the authors' intellectual openness, as, for example, when they state that "young earth creationism, therefore, need not embrace a dogmatic or static biblical hermeneutic," but "must be willing to change and admit error." This seems to be consistent with a truly "open" philosophy of science, one that is open to new information, from whatever source it might come.

Having noted these areas of agreement, I must also state that I find substantial areas in the authors' treatment that are, in my judgment, inadequate both in terms of biblical interpretation and in its understanding of the scientific evidence.

The authors state that they believe that "all basic types of organisms were directly created by God." This broad generalization does not seem to be adequately supported either by a detailed examination of the fossil record or by specific examination of the biblical usage of terms for "create." For example, the extensive fossil evidence relating to the extinct mammal-like reptiles, which display many characteristics intermediate between modern reptiles and mammals, is not addressed.[3] Similarly, there is no attempt to grapple with the significance of the hominid fossil record—with the extensive paleontological record of the *australopithecines*, *Homo erectus*, the Neanderthals, and so on—that indicates that modern *Homo sapiens* did not appear on earth without precursors. In features such as bipedalism, growth in brain capacity, and changes in dentition and skeletal structure, these extinct hominid forms displayed characteristics intermediate between modern man and earlier primates.

The authors' paper seems to presuppose a certain understanding of the biblical concept of creation, rather than basing such a concept on a careful inductive study of the variety of biblical usage in this area. The authors' treatment would have benefited, for example, by a more careful distinction between the concepts of *primary* and *secondary* causes in God's creative activity. The biblical writers can describe God as the *primary* or ultimate

---

[2]Michael J. Behe has powerfully expounded the concept of "irreducible complexity" in *Darwin's Black Box* (New York: Free Press, 1996).

[3]Such evidence is discussed, for example, in Robert L. Carroll, *Vertebrate Paleontology and Evolution* (New York: W. H. Freeman, 1988), and T. S. Kemp, *Mammal-Like Reptiles and the Origin of Mammals* (London: Academic Press, 1982).

cause of a natural process, such as the growth of grass for the cattle (Ps. 104:14), and make no reference at all to the *secondary* causes (such as photosynthesis), which are the focus of attention in modern science but of relatively little interest to the biblical writers. The biblical writers are more concerned with the *results* of God's creative work than with the *material processes* that may have been used; their primary concern is *religious* rather than "scientific," being concerned primarily with the *significance* of natural forms in relationship to the *redemptive* purposes of God. These distinctions allow us to recognize that such "primary cause" descriptions of God's creative activity are not theologically or biblically inconsistent with scientific "secondary cause" descriptions of biological origins that could, in a given case, involve transitional forms between major classes or phyla.

The authors speak of the "creation week," but do not argue exegetically for the literal-day view. Nor do they give serious consideration to other ways of understanding the days of Genesis 1, such as the "framework hypothesis" that sees the six days as a literary framework for God's creative activity.[4] The authors' treatment would have benefited by a more careful examination of the religious and cultural context of the Ancient Near East within which the text of Genesis is most naturally understood. In this context it becomes clear that the primary purpose of the Genesis text is to communicate *theological* truths about the God of Israel against the backdrop of polytheistic deities of the Ancient Near East, rather than to answer the specific, process-oriented questions of the modern scientific age.[5]

In speaking of the "creation week" and a "young earth," the authors refer to organizations such as the Institute for Creation Research, but detailed evidence for a young earth is not discussed. Also, the authors do not seriously grapple with the comprehensive body of scientific evidence for an old earth drawn from astronomy, physics, and sedimentary geology dis-

---

[4]For this view, see Henri Blocher, *In the Beginning: the Opening Chapters of Genesis* (Downers Grove, Ill.: InterVarsity Press, 1984), 39–59, and other authors cited in n. 5 of my "Response to Newman" on page 138 below.

[5]The Ancient Near Eastern context of Genesis is helpfully reviewed in Claus Westermann, *Genesis 1–11: A Commentary*, trans. John J. Scullion (London: SPCK, 1984), esp. 19–46, "Creation in the History of Religions and in the Bible," and in Gerhard F. Hasel, "The Polemic Nature of the Genesis Cosmology," *Evangelical Quarterly* 46 (1974): 81–102.

cussed by authors such as Davis Young, Dan Wonderly, Alan Hayward, and Brent Dalrymple.[6] The authors are to be commended, however, for their admission that recent creationism has "a problem with starlight and the size of the universe," and seem open to further consideration of these issues.

Nelson and Reynolds state that the "curse of Genesis 3:14–19 profoundly affected every aspect of the natural economy." According to their understanding, there was no animal death in the world before the sin of Adam. This point of view is based on a certain interpretation of texts such as Romans 5:12, "sin entered the world through one man [Adam], and death through sin." It is not necessary to understand the text in this way, however. The rest of this verse makes it clear that the apostle Paul is concerned with *human death* as a punishment for sin, not biological death in general: "In this way death came *to all men,* because *all* sinned [in Adam]" (Rom. 5:12b). A proper understanding of this verse does not require us to deny the massive evidence of animal death attested by the fossil remains in the lower sedimentary strata long before the appearance of man. Furthermore, as Alan Hayward has pointed out, the warning given to Adam before the Fall, "when you eat of it you will surely *die*" (Gen. 2:17), could have had little or no meaning if Adam had never observed the death of any plant or animal.[7]

Nelson and Reynolds hold the universal flood view, which of course has a long and honorable history in Christian thought. They do not, however, give serious exegetical attention to texts such as Genesis 41:57 ("All the countries came to Egypt to buy grain from Joseph, because the famine was severe in all the world"); Deuteronomy 2:25; 1 Kings 18:10; 2 Chronicles 9:23 ("All the kings of the earth sought . . . Solomon to hear [his] wisdom"); Acts 2:5 ("from every nation under heaven" [China?]); Colossians 1:23 ("the gospel . . . proclaimed to every creature under heaven"); and so forth, which apparently indicate that global language in some contexts of Scripture can have limited reference, or it can be global *from the perspective and purposes of*

---

[6]Complete references to these authors have been given in n. 4 of my "Response to Newman."

[7]Alan Hayward, *Creation and Evolution* (Minneapolis: Bethany House, 1985), 182. Note also his comments on the Fall and the second law of thermodynamics on pp. 183–84.

the writer.[8] Problems of the geographical distribution of species that bear on the question of the extent of the Flood are not addressed. For example, did kangaroos and platypuses that live in Australia come to the Middle East to board Noah's ark? In Genesis 6:20 the text states that the animals in question would *come* to Noah, but it does not indicate that the animals were *miraculously transported* by God. On this and other issues, the authors' treatment would have benefited by further attention both to questions of biblical interpretation and a broader range of empirical data.

---

[8]For a discussion of these and other biblical and scientific arguments for a localized understanding of the Flood account, see Donald C. Boardman, "Did Noah's Flood Cover the Entire World," in *The Genesis Debate*, ed. Ronald Youngblood (Nashville: Thomas Nelson, 1986), 210–29. In *The Biblical Flood* (Grand Rapids: Eerdmans, 1995), Davis A. Young demonstrates from church history how Christian thinkers in earlier generations have incorporated new scientific information into their understandings of the biblical account of the Flood.

RESPONSE TO PAUL NELSON AND JOHN MARK REYNOLDS

J. P. Moreland

While I lean somewhat strongly toward an old earth creationist view, I cannot shake the idea that the young earth people may be correct. In response to Nelson and Reynolds, I want to give young earth creationists two pieces of advice and close by asking why it is that evolution is so widely embraced with such certainty in a way that goes well beyond what the evidence justifies.

Here's my first piece of advice: All disciplines have theories for which there are external conceptual problems. An external conceptual problem is an intellectual difficulty for a theory in a discipline that has its origin in a field outside that discipline and which tends to disclaim the theory in question. For example, if there are philosophical arguments for the fact that the past could not have been infinite in duration, these present external conceptual problems for scientific models that imply an infinite past. Now it seems to me that one of our intellectual duties as Christians is to interpret Scripture with an eye on solving external conceptual problems.

Suppose we are interpreting some biblical text and we have hermeneutical option A and option B. Suppose further, that on exegetical grounds alone, we compare the text with other portions of Scripture and find that (1) A and B are both plausible, that is, within the bounds of reason exegetically speaking; and (2) A is superior to B. Now suppose further that B harmonizes Scripture with what we have pretty good reason to believe is true outside the Bible, but A flies in the face of these extrabiblical factors. In short, B solves external conceptual problems. Then, in my view, it is hermeneutically permissible to adopt B as the correct interpretation of a text.

I am not advocating that we make our interpretations of Scripture dependent on widely accepted scientific claims. Nelson and Reynolds correctly warn us against this. I am simply arguing that we do need to solve external conceptual problems as one aim of biblical exegesis. Clearly, criteria should be developed for deciding when one has gone too far in allowing extrabiblical data to inform an interpretive option, and Nelson and Reynolds do offer some criteria in this regard. When these criteria are developed, I believe they will include the admonition that we draw the line at the place where B steps outside of what can be justified as intellectually plausible or permissible on exegetical grounds alone apart from the extrabiblical issue. We must also require that the alleged external conceptual problem is highly justified from a rational point of view. When it comes to the age question, I take A and B to be the young and old earth views, respectively. I don't accept theistic evolution as a candidate for B for two reasons. First, I don't think it is plausible on exegetical grounds alone; I think it is an ad hoc attempt to save Genesis from embarrassment in light of evolutionary theory. Second, I don't think the rational justification for evolutionary theory is anywhere near what it needs to be to require us to adjust our interpretations of Scripture. Young earth creationists need to interact with this approach to the problem because it has some merit to it.

My second piece of advice is directed toward young earth attempts to rebut evidence for an old universe on the grounds that creation would have the appearance of age. I am sympathetic to this claim, but those who employ it could help some of us become more convinced of its legitimacy by addressing the following problem. (Nelson and Reynolds do address this issue, but I want to reiterate its importance here and clarify the problem.) Let us distinguish those factors in the initial creation that were functionally necessary to have a Garden of Eden from those that were not. Examples of the former are light from the heavens, full-grown trees, and rivers with some depth to them. Examples of the latter would be rings inside the trees, stars that appear to be traveling away from one another, and certain ratios of uranium, lead, and other elements used to date rocks. Now I can easily see why God would create with an appearance of age those things functionally necessary for Adam and Eve to live their lives (e.g., full-grown trees), but I don't see why he would create nonfunctional factors that indicate an old age. One response would be to argue that all the supposed nonfunctional factors do indeed have a function, even if we don't know what it is at this point. In any case, this is an issue young earth creationists need to address.

So much for my advice. It is sometimes said that young earth creationists embrace their position with a degree of commitment that goes well beyond what the evidence justifies. I doubt this claim, but I am not a sociologist or the son of one, so I leave the matter to others. What I am certain of, however, is that this claim applies to most advocates of evolution. My point here is not about the empirical evidence for evolution. I think this evidence is quite meager. But in any case, even if we grant, for the sake of argument, that there is a decent bit of positive evidence for it, the degree of certainty claimed on its behalf, along with the widespread attitude towards creationists, is quite beyond what is warranted by the evidence alone. What is going on here? At least two things.

First, the monolithic intellectual authority of science, coupled with the belief that special creationism is religion not science, means that evolution is the only view of origins that can claim the backing of reason. In our sensate culture, science and science alone has unqualified intellectual acceptance. On the evening news, when a scientist makes a pronouncement about what causes obesity, crime, or anything else, he is taken to speak ex cathedra. When was the last time you saw a philosopher, theologian, or humanities professor consulted as an intellectual leader in the culture? All supposedly extrascientific beliefs must move to the back of the bus and are relegated to the level of private, subjective opinion.

Now, if two scientific theories are competing for allegiance, then most intellectuals, at least in principle, would be open to the evidence relevant to the issue at hand. But what happens if one rival theory is a scientific one and the other is not considered a scientific theory at all? If we abandon the scientific theory in favor of the nonscientific one, then given the intellectual hegemony of science, this is tantamount to abandoning reason itself. If we can draw a line of demarcation between science and nonscience, a set of necessary and sufficient conditions that form a definition of science, and show that creationism is religion masquerading as science, then the creation-evolution debate turns into a controversy that pits reason against pure subjective belief and opinion. In the infamous creation science trial in Little Rock, Arkansas, in December of 1981, creation science was ruled out of public schools, not because of the weak evidence for it, but because it was judged religion rather than science. Today, in the state of California you cannot discuss creationist theories in science class for the same reason.

Space forbids me to present reasons why almost all philosophers of science, atheist and Christian alike, agree that creation science is at least a science and not a religious view, regardless of what is to be said about the empirical evidence for or against it.[1] Suffice it to say that philosophical naturalists are currently in control of who sets the rules for what counts as science. The bottom line is this: Philosophical naturalism is used to argue both that evolution is science and creation science is religion, and that reason is to be identified with science. Thus, the empirical evidence for or against evolution is just not the issue when it comes to explaining why so many give the theory unqualified allegiance.

There is a second reason for the current overbelief in evolution: it functions as a myth for secularists. By myth I do not mean something false, though I believe evolution to be that, but rather a story of who we are and how we got here that serves as a guide for life. Evolutionist Richard Dawkins said that evolution made the world safe for atheists because it supposedly did away with the design argument for God's existence. In graduate school, I once had a professor say that evolution was a view he embraced religiously because it implied for him that he could do anything he wanted. Why? The professor went on to say that, given that there is no God and that evolution is how we got here, there is no set purpose for life given to us, no objective right and wrong, no punishment after death, so one can live for himself in

---

[1] I have presented these arguments in *The Creation Hypothesis* (Downers Grove, Ill.: InterVarsity Press, 1994) and in *Christianity and the Nature of Science* (Grand Rapids: Baker, 1989).

this life any way he wants. Serial killer Jeffrey Dahmer made the same statement on national TV. Dahmer said that naturalistic evolution implied that we all came from slime and will return to slime. So why should he resist deeply felt tendencies to kill, given that we have no objective purpose or value and there is no punishment after death? I am not here arguing that secularists cannot find grounds for objective purpose and value in their naturalistic worldview, though I believe that to be the case. I am simply pointing out that evolution functions as an egoistic myth for many intellectuals who have absolutized freedom, understood as the right to do anything I want. Philosophical naturalists *want* evolution to be true because it provides justification for their lifestyle choices.[2]

For these two reasons—the identification of evolution as the only option on origins that claims the support of reason and the function of evolution as a convenient myth for a secular lifestyle—the widespread overcommitment to evolution is not primarily a matter of evidence. That is why people react to creationism with hatred, disgust, and loathing, instead of responding to creationist arguments with calm but open-minded counterarguments. This situation is tragic, because it has produced a cultural logjam in which philosophical naturalism is sustained as our source of cultural authority, protected from serious intellectual criticism and scrutiny.

For Christians, there is a lesson to be learned from all this and an application to be followed. The lesson is this: The debate about creation and evolution is not primarily one about how to interpret certain passages in Genesis, though it does include that. Rather, it is primarily about the adequacy of philosophical naturalism as a worldview and the hegemony of science as a cognitive authority that relegates religion to private opinion and presuppositional faith. The application is this: Believers owe it to themselves and the church to read works that present a well-reasoned alternative to evolution and to keep an eye on the broader implications of taking theistic evolution as a via media.

---

[2]Thomas Nagel, *The Last Word* (New York: Oxford University Press, 1997), 130–31.

RESPONSE TO PAUL NELSON AND JOHN MARK REYNOLDS

Vern S. Poythress

Dr. Nelson and Dr. Reynolds's essay makes a number of valuable points. It puts emphasis on being open to a variety of scientific models. It warns against getting swept up into naturalistic assumptions and foreclosing other kinds of explanations. It calls on us to do justice to the teachings of the Bible, and not merely to assume that the Bible rather than science must be reinterpreted when the two appear to conflict. It candidly admits that currently "young earth" theories are not as scientifically attractive as an "old earth" approach. All this is helpful.

The essay's greatest weakness, however, lies in how briefly it actually discusses the Bible. More discussion is needed on this, since it is crucial that young earth theorists make sure whether the Bible unequivocally supports a young earth. In particular, the central issue of how to interpret the Bible needs fuller discussion. In its brevity the essay tends to polarize interpretation into two alternatives. According to "mythological" interpretation, the text sets forth some theological truth but with no real, factual history behind it. By contrast, in "natural interpretation" the text talks in a straightforward way about what really happened. Unfortunately, this idea of natural interpretation fails to distinguish two quite different approaches: grammatical-historical interpretation, in which one attends closely to the Bible's actual meanings within the Ancient Near East; and naive-modern interpretation, in which one reads the text only against the background of one's own modern world and life.

Against the background of the Ancient Near East, the passages in Genesis 1–11 describe real events. They are not myth. But God is primarily interested in attacking polytheism and pagan myths. He has spared us from technical details that the Israelites did not need to know and would not understand.

Consider an illustration. The essay says that "the Bible seems to support a still earth on the most natural reading of the text." This supposed "natural reading" is in fact a *misreading* of the actual intent of the text. The Bible's language concerning the earth's not moving (e.g., Ps. 93:1) is "phenomenal language," the language of appearances. It describes what any human being can see: the earth remains solidly underfoot. Just look beneath you! Psalm 93:1 is a true statement, but is misinterpreted when we imagine that God is providing the Israelites with the technical foundation for a modern scientific-theoretical model of the solar system.

For a long time careful interpreters of the Bible have recognized the *ordinariness* of God's descriptions of the created world. In 1554 John Calvin said concerning Genesis 1:14, "It must be remembered, that Moses does not speak with philosophical acuteness on occult [secret] mysteries, but relates those things which are everywhere observed, even by the uncultivated, and which are in common use."[1] Bernard Ramm in *The Christian View of Science and Scripture* discusses the same issue at length.[2]

Once we fully take into account the character of biblical language, the supposed biblical support for a young earth becomes questionable. For example, by adding up the dates from the genealogies in Genesis 5 and 11 we can obtain an estimate for the date of creation. But it is well known that genealogies in the Bible may contain gaps (e.g., Matt. 1:1–16). Thus the Bible simply does not tell us that we can obtain an accurate dating for creation by simple addition.

Consider next the Bible's account of the Flood in Genesis 7:17–24. Sound interpretation shows that the text is describing

---

[1]John Calvin, *Commentaries on the First Book of Moses Called Genesis* (Grand Rapids: Eerdmans, n.d.), 1:84.

[2]Bernard Ramm, *The Christian View of Science and Scripture* (Grand Rapids: Eerdmans, 1954), 65–80.

real events and a real person, Noah. It is not myth. But the text describes things as they would appear to a human observer like Noah. Everything within range of human observation was covered with water, and all the animals within range died.

"All the high mountains under the entire heavens were covered" (Gen. 7:19). The reader bent on naive-modern interpretation rushes to conclude that the water must have covered the entire globe. But note the word "under." From the point of view of modern *technical* astronomical language, the globe of the earth is not "under" space, but within it. The word "under" in Genesis 7:19 confirms the ordinary character of biblical language.

So what does the passage actually teach? The description addresses what Noah could have seen. The sky is above him. All the mountains that he can see in any direction are under it. All these mountains are covered with "waters." Were all the "waters" liquid, or did some take the form of snow or ice on the high mountains? God does not supply us with technical detail, because the point on which the passage focuses is the universality of the divine judgment, within the scope of what Noah could see. The Bible simply does not say whether the Flood covered the entire globe.[3]

One must also look carefully at God's description of creation in Genesis 1–2. Many modern readers naively conclude that the acts of creation must have taken six twenty-four-hour days. But careful grammatical-historical interpretation moves us in another direction. A technical scientific description of the timing and the mechanisms is not the primary focus of Genesis 1–2. If we want nevertheless to probe for hints concerning exact timing, we run into difficulties, as Meredith G. Kline has shown.[4]

First, the seventh day, the day of God's rest (Gen. 2:2–3), goes on forever. Though God continues to act in providence and in salvation, his acts of *creating* are finished forever. But if the seventh day is God's eternal rest, the other six days are also *God's* days, not simply ours; we cannot naively deduce that they must be twenty-four hours long. Second, in creating the Garden of

---

[3]The waters covered the "earth" and the "dry land"—that is, what Noah perceived underfoot. No word corresponding to our modern theoretical idea of the "globe" occurs in the passage.

[4]Meredith G. Kline, "Space and Time in the Genesis Cosmogony," *Perspectives on Science and Christian Faith* 48 (1996): 2–15.

Eden, God caused trees to grow up (Gen. 2:9). The specific language indicates not creation in a moment, but rather a process of growth. God's agricultural work in creating the Garden is a foundation and model for man's agricultural work (Gen. 2:15; 3:17–19). But the model is analogous to—rather than identical with—man's work. How long does it take? For a man it would take years. Does God proceed by the same timescale or a different scale? The text does not answer. Similarly, God's week of work and rest in Genesis 1:1–2:3 is a model analogous to—rather than identical with—man's week (Exod. 20:11). We cannot safely calculate length. Finally, God made the heavenly bodies on the fourth day to "serve as signs to mark seasons and days and years" (Gen. 1:14). These bodies are a kind of mark or standard so that human beings can identify days and years. We have no business trying to calculate the length of days independent of this pattern. Trying to give a timing to the first three days ignores God's own provision for us. Instead, we should take Genesis 1:14 seriously. The role of the sun in governing the day (Gen. 1:16) should make us wonder about the first three days. It suggests that the arrangement of acts in Genesis 1 may be partly topical and not purely chronological.[5]

In actuality, then, several clues in the text itself show us that Genesis 1 provides a framework of seven days of God's action as the basis for human action. But the details caution us against postulating an identity, rather than an analogy, between God's days and ours.

Some lesser problems with Nelson and Reynolds's essay deserve mention. Like the two other essays, this essay also fails to achieve theological clarity about what it means for God to act "directly." In a sense, God directly makes the grass grow (Ps. 104:14), as I explain in my other two responses. We must beware of thinking that God's action is limited to what is exceptional, or that *only* exceptional events provide evidence for his presence.[6]

The essay hopes for better days for "young earth" scientific theories. Of course, such a turn of events is always theoretically

---

[5]See also David Sterchi, "Does Genesis 1 Provide a Chronological Sequence?" *Journal of the Evangelical Theological Society* 39 (1996): 429–536.

[6]Romans 1:18–20 and Acts 14:15–17 indicate that every form of God's rule over the world, not merely a surprising exceptional act, leaves, as Nelson and Reynolds said, "unmistakable evidence of his existence."

possible. But people have been working on young earth flood theories since at least the eighteenth century,[7] and, if anything, the situation has become much more difficult for them with advances in astronomy and geology (including radioactive dating and plate tectonics).

[7]Davis A. Young, *The Biblical Flood: A Case Study of the Church's Response to Extrabiblical Evidence* (Grand Rapids: Eerdmans, 1995).

CONCLUSION

Paul Nelson and John Mark Reynolds

> From the facts above enumerated it is clear that certain fishes come spontaneously into existence, not being derived from eggs or from copulation. Such fish as are neither oviparous nor viviparous arise all either from mud or from sand and from decayed matter that rises thence as scum; for instance the so-called froth of the small fry comes out of sandy ground.[1]

Modern readers find this quotation from Aristotle amusing. How could anyone have believed such things? The vast majority of thinkers, however, agreed with Aristotle. Spontaneous generation seemed to fit the data. To suggest otherwise was to invite ridicule. Theologians, such as Albertus Magnus, carefully crafted such ideas into their theology.[2] Francis Bacon, often seen as critical in the development of experimentation in science, accepted the concept in his *Novum Organum*.[3] Aristotle himself listed scores of examples from nature to support his view. Despite contrary experimental data as early as the seventeenth century, it was not until the work of Louis Pasteur that such ideas were finally laid to rest.[4]

---

[1]Aristotle, *History of the Animals*, in *The Complete Works of Aristotle*, ed. Jonathan Barnes (Princeton: Princeton University Press, 1991), 569a10.

[2]A. C. Crombie, *Medieval and Early Modern Science* (Cambridge: Harvard University Press, 1963), 1:154.

[3]Francis Bacon, *Novum Organum*, trans. and ed. Peter Urbach and John Gibson (Chicago: Open Court, 1996), 160.

[4]Urbach and Gibson point out (p. 160) that at least insect spontaneous generation had been ruled out by Francesco Redi in the seventeenth century. Scientists

Of course, everyone now concedes that this scientific "truth" is false. We are even tempted to say that it was obviously false, and only a perverse sort of intellectual blindness prevented people from seeing that this was so. At the time, however, needful changes in thinking were impeded by an unreasoning commitment to what "seemed to be so." Revolutions in thought come slowly, sometimes too slowly. It is always safest to agree with the spirit of the age. But it is not always best.

At the same time, it is also important to remember that quacks and lunatics often fancy themselves on the edge of a "revolution" in thought. "If only the mainstream would listen," says the owner of "Bigfoot" relics or pieces of Noah's ark, "it would all be clear to them." But of course there is not time in the day or rational reason for the culture to listen to such deluded souls. They are on the fringe and it is the sign of the health of the academic culture that they remain on the fringe. Cultures in decay often show an unhealthy interest in such marginalia.[5]

This is the problem that troubles any thoughtful young earth creationist. Are they on the edge of a new understanding? Are they revolutionaries for a better world? Or are they certified members of intellectual bedlam? Of course the reaction of the culture cannot tell us the answer. True innovators are often ridiculed or worse, but then so are the true fanatics. The key is to figure out who fits into what category.

Both of us are sure that one cannot solve the problem by a simple weighing of the evidence. The most plausible case at any moment is usually going to be that of the establishment. In our own era the establishment is most comfortable with naturalism and its philosophic relatives. Thousands of scientists, theologians, and philosophers have spent the last one hundred years, with almost the full backing of the Western cultural apparatus, accumulating data filtered through this point of view. This is not to postulate some grand conspiracy or to trash professional academics. We merely accept the commonsense dictum: "Ideas have consequences. These ideas shape what people see." When people believed that women were, in the words of Dorothy Say-

---

seemingly just went on to other candidates for spontaneous generation. The idea itself was so deeply ingrained that it died a slow death.

[5]One thinks of the interest in the paranormal in Russian society before the disaster of 1917.

ers, "not quite human," all sorts of data were interpreted as fitting that idea. Later, when attitudes began to shift, the evidence did not change and very little new evidence was gathered, but the interpretation of the evidence certainly did shift.

We believe that there are at least three good reasons for taking young earth creationism seriously and for not relegating it to the same shelf as speculations about the lost city of Atlantis. First, young earth creationism has grown and developed intellectually over time. One need only compare seminal works such as the *Genesis Flood,* published in 1961, with the papers coming out of the last International Conference on Creationism in 1998. These papers manifest a great increase in the sophistication and rigor of the arguments being deployed. If young earth creationism is a pseudoscience, it is the only one we are aware of that has grown in this manner. Usually a pseudoscience begins with some semicredible figure. He or she writes a book explaining the "new science." From that point on, cheap pop versions of the "new science" are deployed. The disciples rarely surpass the master. Of course, in real science the students *all* surpass the master in terms of the building up of knowledge. The same process can be seen in young earth creationism as it matures. Kurt Wise, a Harvard-trained paleontologist with young earth views, has a stronger knowledge of the contemporary geological record than many of his old earth critics.[6]

Second, young earth creationism has been the overwhelming view of the traditional church. This does not make it true, but we are wary of new readings of the text that conveniently fit the "spirit of the age" (*Zeitgeist*). A Bible written by God so that only the modern reader can fully understand its message strikes us as implausible. The Fathers from the first century forward overwhelmingly took a young earth, global-flood view. Catholic and Orthodox believers, who make up more than two-thirds of Christendom, take this tradition very seriously. Simply discarding the views of the Fathers is not an option for any thoughtful Christian.[7] We are also fairly skeptical of many of the assumptions,

---

[6]See Kurt Wise, "The Origin of Life's Major Groups," in *The Creation Hypothesis,* ed. J. P. Moreland (Downers Grove, Ill.: InterVarsity Press, 1994), a splendid book.

[7]One need only note the great interest in Catholic, Orthodox, and Protestant circles in the new *Ancient Christian Commentary on Scripture* (Downers Grove, Ill.: InterVarsity Press, 1998).

themselves naturalistic, behind some modern biblical scholarship.[8] It may be decades after the growth of nonnaturalistic science before theology itself is free from the intellectual constraints of naturalism.

Third, young earth creationism is intellectually exciting. It has some empirical evidence that is already in its favor[9] and an exciting, potential, research program. For example, in the 1994 *Proceedings of the International Conference on Creationism,* six creationist scientists, all with relevant terminal degrees, presented a paper entitled "Catastrophic Plate Tectonics: A Global Flood Model of Earth History." This paper provides a new theoretical way of understanding the flood of Noah and its impact on the geological record. It solves many problems, while providing a huge amount of room for future research. Whether this particular theory is successful is not important. The pivotal fact is that *good and interesting science can now be done in a young earth framework.* We believe these reasons alone are sufficient to allow a reasonable person to call himself provisionally "young earth." Of course, like any other thoughtful person, he should also be open to other possibilities. This open attitude is not incompatible with taking a particular approach to thinking and research.

Our critics have asked many good questions about specific scientific and theological areas. They deserve thoughtful answers. We intentionally, however, choose to refrain from dealing with such details in this concluding essay, important though they may be. There are two reasons for this. First, young earth creationism is generally underdeveloped. It is not ready for such a specific interchange. Second, we believe it would distract from the main point. *We agree with our respondents on the essential philosophical issue. If an open philosophy of science triumphs, and young earth creationists are allowed to do their research as full members of the academy, we will have gained the critical point.* The details of Genesis and biblical chronology are important, but they are less important than this singular fact: If theism is true, any reason-

---

[8]See Eta Linnemann, *Historical Criticism of the Bible: Methodology or Ideology?,* trans. Robert W. Yarbrough (Grand Rapids: Baker, 1990).

[9]See the collected *Proceedings of the International Conference on Creationism.* These volumes are a rich resource for specific arguments. The papers are at times uneven in quality, especially in the first two conferences, but they show growth and promise for the movement.

able philosophy of science must be open to a God who works in scientifically detectable ways.

In many cases, young earth creationists would need decades of fully funded research just to begin to get a grasp on a new way of looking at the mountain of current data. Skeptics of young earth creationism sometimes claim that we, too, have had centuries to work on these problems, but this is false. Young earth creationism failed to answer the initial geological (prior to Darwin) and biological (after Darwin) challenges. With the advent of each challenge, almost all scientists, philosophers, and theologians of note "switched sides." This was due in part to a commitment to naturalistic methodology that made any non-supernatural answer preferable, even for theists, to any theistic one. *The problem was philosophical* and not a matter of "evidence."

Scientists like Michael Behe in *Darwin's Black Box*[10] are beginning to demonstrate that philosophical naturalism may be retarding the advancement of science. In biology and psychology, it may be that the old nineteenth-century ideas are finally being recognized as slowing progress rather than aiding it. Philosophers like William Dembski are developing new ways of understanding design. His book *The Design Inference*[11] is an explosive book in this area. These revolutionary ways of thinking may help liberate biology and psychology. It is too soon to know, but the signs are promising. The alternative is to play it safe and allow past failures to hold back intellectual growth. Science, philosophy, and theology cannot grow by such overly rigid conservatism. The almost obsessive fear of a God-of-the-gaps argument in some religious thinkers is a manifestation of such intellectual timidity.

Whatever the truth of the matter may be in regard to biblical history, we are, perhaps of all ages, least likely to find it. Nothing about the education of most moderns leaves them disposed to be sympathetic to traditional readings of the biblical text. The traditional Christian faces the intellectual demand that he "harmonize" his unfashionable faith with the demands of a culture firmly fixed in the opposite direction. The almost overwhelming temptation is to "trim." Suddenly, new ways of reading the text of Scripture are discovered, which to no one's

---

[10]Michael Behe, *Darwin's Black Box* (New York: Free Press, 1996).

[11]William Dembski, *The Design Inference: Eliminating Chance Through Small Probabilities*, (Cambridge: Cambridge University Press, 1998).

surprise allow for accommodation between at least some of the reigning paradigms and traditional religion. We are suspicious of such easy discoveries.

Our advice, therefore, is to leave the issues of biblical chronology and history to a saner period. Christians should unite in rooting out the tedious and unfruitful grip of naturalism, methodological and otherwise, on learning. At this moment in history, we need not be concerned with differences between advocates of intelligent design on such issues. Our reviewers are not comfortable with young earth creationism in its current state of development. We do not blame them for this discomfort. They seem willing to allow for an opportunity at a more robust theory on the part of those "young earthers" willing to make the attempt. We are thankful for this tolerance. Our agreements with our reviewers far outweigh the importance of our disagreements.

It is obvious that a person who is generally committed to a traditional understanding of Christianity can be "old earth." Persons interested in arguments on either side of this issue can refer to the books listed in our bibliography and footnotes. Our disagreements on these points should not distract from the main topic. Philosophical naturalism is retarding science, philosophy, and theology. It seems to both of us that our reviewers agree in finding such a situation intolerable. *To fail to unify with such people of goodwill in the assault on naturalism would not just be foolish; it would be intellectual treason.* There is no reasonable chance that a society forcibly wedded to naturalism will be interested in the young earth project. When the intellectual climate is different, the time will have come to explore these important issues.

The key thing is to oppose any sort of attempt to accommodate theism and naturalism. The "theistic naturalism" that results is intellectually impotent and culturally marginal. It apes the language both of traditional religion and science. It creates its own language, discovering new meanings in old words. It creates a whole new overly elaborate vocabulary when standard language threatens too much clarity. Generally, it serves as a stopgap for persons who confuse too much intellectual conservatism with rigorous thought.

Theistic evolutionists have little or no cultural or intellectual power. They do not shape the discussion. They are generally only invited to speak as some sort of "safe" religious spokespeople. Their appearance at otherwise dominantly secular events lulls the religious community into a contented intellectual slumber.

Calls to find God's action in the "creaturely capacities" of an evolving natural world only have a new vocabulary. Their core appeal is not new. In many ways, such more sophisticated writings echo the advice given in a little book written by a Presbyterian minister early in this century. He appeals for "some sympathetic, intelligent help in bridging the chasm between Professor's Science and Mother's religion." What is the answer? It is to realize that evolution is not the foe. He notes, "I have wandered in the labyrinth of Evolution and have enjoyed something of the thrill of its immense distances and endless dimensions, and still have kept my faith in the God of the first chapter of Genesis." He sees God's hand in the work of evolution and in his summation he delivers the following poem to his reader:

> A firemist and a planet
> A crystal and a cell
> A jelly-fish and a Saurian
> And caves where cave-men dwell
> Then a sense of law and beauty,
> And a face turned from the clod—
> Some call it Evolution
> And others call it God.

Whatever the merits of this poem as literature, it is an utter failure as intellectual advice. The culture recognizes that almost any two positions can, with sufficient verbal skill, be made compatible. The question is, What view is most plausible given the things we believe? If naturalism is the way the world works, and produces the methodology that can answer all the big questions, then naturalism will carry the day. It may be that with care theism can be made logically compatible with the triumph of methodological naturalism, but most people will sensibly turn away. The academic and popular culture kept "Professor's Science" and sadly discarded "Mother's religion." That is just the way sensible people think. Whatever his or her view of Genesis, therefore, the traditional Christian must embrace a robust and plausible theism that dares to test its claims against science and history. We believe that Phillip E. Johnson has been instrumental in creating a new way of looking at the old religion and science program. This helpful vision unites the traditionally religious; it does not divide them. It is our hope that old and young earth creationists can set aside their differences to implement that vision.

Chapter Two

PROGRESSIVE CREATIONISM

Robert C. Newman

PROGRESSIVE CREATIONISM

("Old Earth Creationism")

Robert C. Newman

1. OVERALL POSITION

Personal Position on the Creation-Evolution Controversy

My position on the creation-evolution controversy is that I am an old earth creationist. As an *old earth* creationist I understand that the earth and the universe were created far more than just a few thousand years ago as has been the traditional belief among Christians. Rather I think the earth is some four or five billion years old and the universe some ten to twenty billion years old.

As an old earth *creationist* I believe that unguided evolution is not capable of producing the features we see in our universe—not the universe itself, life, its actual variety, not humankind. Nor do I think that God-guided evolution is the way God chose to create, at least not to produce the large-scale differences between the various plants and animals, nor to make humans. Presumably God is capable of creating everything we see either by means of miracles in just a few days (even no time at all!) or by guiding purely natural processes over a long period of time. But I don't think the biblical or scientific evidence we have suggests that he used either of these means exclusively. Instead, it seems to me that God used some combination of supernatural intervention and providential guidance to construct the universe. Perhaps he did this so that the universe would be of such a sort as to display design and structure far surpassing its own innate capabilities, thus sending us a message about the existence and character of our Creator (Ps 19:1–4; Rom 1:19–20).

This old earth position is also sometimes called "progressive creationism." This is not because we think ourselves to be progressive while young earth creationists are "reactionary." (We ought never to look down on people for trying to hold firmly to what they understand God has said.) Rather it's because we think God's activity in creation occurred in a *progression*—a number of steps over a long period of time in which God established and perfected each level of the environment before he added a higher level that rests (so to speak) upon the preceding levels.

There are a number of varieties of old earth creationism, just as there are varieties of young earth creationism and theistic evolution. A sort of intermediate position between young earth and old earth is the "gap theory," which sees God's original creation of the universe and earth (taking ages) mentioned in Genesis 1:1 ("in the beginning God created the heavens and the earth"), followed by the destruction of the earth's habitat (perhaps due to Satan's rebellion) in Genesis 1:2 ("the earth was [or possibly 'became'] formless and empty"). The rest of the Genesis account then describes the restoration of the earth just a few thousand years ago in six literal days. In this view, popularized in the old *Scofield Reference Bible*, geologists are looking at the original creation and Genesis is looking at the restoration.

Most varieties of old earth creationism, however, see the Genesis account and the data of cosmology and geology as referring to the same events—the creation of the universe, earth, and their contents. Variations within this position commonly concern how the days of Genesis are to be understood: Are they long periods of time (day-age view), literal days separated by long periods (intermittent-day view), or are the days a literary device rather than an actual chronological sequence (framework hypothesis)? Each of these views in turn has subvarieties with different correlations between features in the biblical text and phenomena in nature, including the question of the antiquity and unity of the human race.

Some proponents of one or the other of these old earth creation schemes include theologians Charles Hodge, Bernard Ramm, and Wayne Grudem; lawyers William Jennings Bryan and Phillip Johnson; geologists Davis Young and Daniel Wonderly; biologist Pattle Pun; chemist Russell Maatman; physicist Alan Hayward; astronomers E. W. Maunder and Hugh Ross; and Old Testament scholars William Henry Green and Gleason Archer, to name a few.[1]

My own view is a variety of the intermittent-day type.[2] After God formed the heavens and the body of the earth in the beginning, each successive day opens a new creative period—day 1 starts the formation of atmosphere and ocean; day 2, the formation of dry land and vegetation; day 3, the oxygenation and clearing of the atmosphere; day 4, the formation of air and sea animals; day 5, the land animals and human beings; and day 6, the formation of redeemed humanity. The seventh day (still future) will open God's eternal sabbath rest, with his people enjoying the new heavens and new earth.

In this sort of scheme, we can get a very nice correlation between the creation account in Genesis and a reasonable model for the earth's origin as commonly proposed by astronomy and geology. Apparently, the narrative is presented so that we readers are observing the events of creation as they unfold around us, as though we are at ground level (once the planet has been formed), rather than imagining we are watching everything from some vantage point out in space.[3] The story goes like this: The earth (with the sun and other planets) was once a shapeless, empty gas cloud. As it contracted under its own gravity, it became dark within (and so to the reader, dark everywhere

---

[1]Some old earth creationist works include Alan Hayward, *Creation and Evolution* (Minneapolis: Bethany House, 1995); Robert C. Newman and Herman J. Eckelmann, Jr., *Genesis One and the Origin of the Earth* (Downers, Grove, Ill.: InterVarsity Press, 1977; Grand Rapids: Baker, 1981; Hatfield, Pa.: Interdisciplinary Biblical Research Institute, 1988); Pattle P. T. Pun, *Evolution: Nature and Scripture in Conflict?* (Grand Rapids: Zondervan, 1982); Hugh Ross, *The Fingerprint of God,* 2d ed. (Orange, Calif.: Promise, 1991); John L. Wiester, *The Genesis Connection* (Nashville: Thomas Nelson, 1983; Hatfield, Pa.: Interdisciplinary Biblical Research Institute, 1992).

[2]Newman and Eckelmann, *Genesis One.*

[3]Dallas E. Cain, "Hindsight Translation of Genesis One," *IBRI Research Report* 43 (1996).

around). Then the whole cloud began to glow (the observer sees light everywhere). The planetary material was pushed out of the cloud and formed up into a rotating planet, with day on the sun side and night on the other side (the observer sees light separated from darkness, the light called "day" and the darkness "night"). The earth's atmosphere was produced from within the planet, separating its waters into surface and atmospheric; the plates making up the crust moved about to open up ocean basins and provide dry land. Plant life appeared and removed carbon dioxide from the atmosphere, lowering earth's temperature, providing oxygen for animal life, and clearing the sky so that the sun, moon, and stars became visible to an observer on the earth's surface. The various forms of animal life appeared on the earth. Finally, human beings were created.

This match between Bible and science would really be quite surprising if the Bible were merely ancient guesswork or made-up stories. But the fit between them is just the sort of thing we might expect if the God who created the universe was also behind the Bible.

Details of this particular exposition aside, why do I think some sort of old earth creation is a better model of origins than atheistic evolution, theistic evolution, or young earth creation? My answers follow below.

Over against young earth creationism, numerous strong scientific evidences (and a few biblical hints) indicate that the earth and universe are very old. For example, take light. As we look out into the sky at night, we can see objects that give every appearance of being many light-years away from us, so that their light began to travel from them to us many years ago. The bright star Sirius, for example, is about twelve light-years away, and the light we now see from it is twelve years old. The Andromeda galaxy appears to be some two million light-years away, so its light would have left it two million years ago. The most distant galaxies and quasars we can see seem to be over ten billion light-years away, which suggests that the universe is at least that old.

Young earth creationists have taken several different tacks to avoid this conclusion. Some think the universe is really quite small, so that only a few years are necessary for light to cross it. Others claim the speed of light was much faster shortly after creation than it is now, so that light from distant objects got here right away. Still others claim that the light we see from distant objects was created on the way, so that we have never actually seen light that left objects more than about ten thousand light-years away.

All these responses seem to face overwhelming problems.[4] If the universe were really quite small physically, then the very dim stars and galaxies we see in our telescopes would also be quite small—too small for gravity to hold them together at their high temperatures. If, instead, the speed of light was nearly infinite at creation, and hundreds to thousands of times faster when Abraham was alive than it is now—then by Einstein's famous formula for the equivalence of mass and energy ($E=mc^2$), the term c^2 (the speed of light times itself) would have been tens of thousands to millions of times larger—so that the sun, in converting a little of its mass to energy, would have fried everyone living on the earth; alternatively, if we make the energy (E) constant, then masses (m) back in Abraham's time would have been so small that the earth's gravity would not have been able to hold on to its atmosphere or even its people!

The most common young earth response is the third alternative mentioned above, namely, that the light from the distant stars was created already on its way to us, so that we could see the stars immediately after they were created, even though there had not been enough time for light to come all the way from them to us. But notice the problem that this produces: when we look at the star Sirius we see what it was doing twelve years ago; when we look at the Andromeda galaxy, we see what it would have been doing two million years ago if it had existed then, but it didn't, so we are really seeing a continuous stream of events that never occurred—fictitious history! As most of the universe is more than ten thousand light-years away, most of the events revealed by light coming from space would be fictional. Since the Bible tells us that God cannot lie, I prefer to interpret nature so as to avoid having God give us fictitious information.

Limitations of space permit me only to briefly mention a few other evidences for an old earth and universe. Some comes to us by calculating the ages of both earthbound, lunar (and perhaps martian) rocks using various radioactive decay processes;

---

[4]Robert C. Newman, "Scientific and Religious Aspects of the Origins Debate," *Perspectives on Science and Christian Faith* 47 (1995): 164–75.

these give ages for various events in the history of these rocks ranging back to a few billion years. Besides this, we have numerous large rock formations on earth that give every evidence of having once been molten but that would not have had time to cool to their present temperatures if the earth were only some thousands of years old. Similarly, calculations of how stars grow old show that some of them are relatively young, but most are a few to many billions of years old. The planets of our solar system have numerous craters in various stages of erosion, sometimes overlapping one another, which testify to a period of several billion years in which the planets were bombarded by meteors. All of this points to an earth and universe far older than a few thousand years.

Although the Bible does not explicitly tell us that the earth is either old or young, a number of biblical hints suggest that it is more than a few thousand years old and that it is much older than the human race. For instance, apart from the Pentateuch, Psalm 90 is the only passage that tells us it is written by Moses. And this is the very psalm that says that God views a thousand years as we would view a day or even a few hours of the night (Ps. 90:4). The apostle John tells us that already in the first century A.D. the "last hour" had come (1 John 2:18), yet that last hour has now lasted nearly two thousand years! What sort of timescale are Moses, John, and God using? Something that allows for ages of earth history?

In the book of Revelation, the apostle tells us that the end of the age will feature an earthquake worse than any that has occurred "since man has been on earth" (Rev. 16:18), which sounds like there might have been bigger quakes before humans were around; this is what geology says, too. Psalm 102:25–26 tells us the heavens will "wear out like a garment," suggesting that they have lasted long enough to age noticeably, a feature more characteristic of billions of years than of thousands, given what we know about stellar-aging processes. None of this gives us numbers for the age of the earth, or even *proves* it is old. But it should make us cautious about climbing out on the limb that it is only a few thousand years old, especially in the face of the scientific data.

Christians all too easily have tended to "overinterpret" Bible passages when other data are lacking, and that sometimes in spite of contrary data. Consider, for instance, the belief that three wise men visited the baby Jesus, when the Bible gives no such number, or that Methuselah was the oldest man who ever lived, when the Bible merely records his age at death and says nothing about whether anyone else ever lived longer.

Speaking of death, young earth creationists maintain that there was no death of any sort (or at least no animal death) before the sin of Adam and Eve. Since the fossil record clearly contains multitudes of dead animals, these must have died sometime after Adam sinned, perhaps during the Flood. Old earth creationists respond that denying any sort of animal death before the Fall is another example of "overinterpreting" the biblical account. Nothing is said one way or the other about animal death in the Genesis account. And the claim that Romans 5 teaches death entered the world through sin is correct, but the context is clearly referring to human death, not animal death. One of the watershed issues dividing young earth and old earth creationists is when animal death first occurred.

Related to this, there are serious problems with the usual young earth explanation for the geologic record—that it was nearly all laid down in the one-year flood during Noah's time. On the contrary, there are many layers in the midst of the geologic record that apparently took much longer than this to form.

For one thing, there are more fossils than this model can explain. If we imagine that nearly all the fossils known to exist were laid down in the Flood, then the plants and animals they represent must all have been alive at the same time; there are so many of them that they must have been crawling over each other many feet deep on its surface! There are also numerous deeply buried layers of rock containing potholes and caverns, showing that the rock from which they were carved was already solid enough to be eroded into vertical (and even undercut) slopes before the new sediment (which fills them) was laid down. The presence of very fragile (but uncrushed) fossils also shows that the sediment containing them was solidified into rock before thick layers of additional sediment were laid on top. Some layered deposits give every evidence of being annual layers, whether by a sequence of salts precipitated from seawater in a tropical bay by evaporation, or by sand and mud laid down seasonally in a freshwater lake by rain and melting snow. Many of these deposits have hundreds of thousands of layers, and some have millions.

Old earth creationists differ among themselves on the extent of Noah's flood. Some, such as James Montgomery Boice and Daniel Wonderly, believe that the water covered the entire earth, but that it did not lay down any large fraction of the geologic strata. Others, such as Frederick Filby and I, favor a flood of limited but very large extent, perhaps filling one of the basins surrounding Eastern Turkey, the traditional site of the mountains of Ararat, such as the Mediterranean, Black Sea, or Caspian basins. Still others favor a Mesopotamian river flood, though it is hard to see how the floodwaters could take so long to recede on this last view.

One of the most striking features of earth's geology is continental drift—a slow movement of the large plates that make up its crust. Today this motion can be measured directly using space-age technology, and is typically an inch or so per year. This rate of motion gives the same ages for the separation of the various continents as do radioactive decay dates for when their geologic strata started to diverge. It also agrees with the dating of magnetic reversals in the new rock formed by magma seeping up where the plates are breaking apart. Likewise, it agrees with the depth and radioactive ages of the ocean-floor ooze deposited on the new crust as it moves away from its place of origin. Here we have the convergent testimony of several diverse witnesses agreeing on an old earth. Young earth creationists try to explain these away by several ad hoc hypotheses, namely, that continental drift right after the Flood was very fast, but it has since slowed down enormously; that the earth's magnetic field oscillated rapidly only *during* the year of the Flood but not since, and so forth. So if we don't try to dismiss the geologic record as fictitious history, it is telling us that the earth is very old.[5]

Yet admitting an old earth does not deliver us into the hands of atheistic evolution. Far from it!

The growing body of evidence from cosmology points steadily to a universe that had a definite beginning, in spite of strenuous attempts to avoid this over the past century by postulating static universes, recycling universes, and currently an

---

[5]One of the best collections of problems with young earth creationism is that by Alan Hayward, *Creation and Evolution* (Minneapolis: Bethany House, 1995), 69–157.

infinite universe in which our universe is only a small bubble.[6] As new data have continued to come in, the spins that some scientists have put on it to avoid belief in a Creator have been successively more and more quirky.

In recent years, it has become apparent that our whole universe is very "finely tuned"—that many of its features need to have just the values they do in order for life to exist. Slight changes in the strength of any of the four basic forces, the expansion speed of the universe, or the character of specific atomic elements would render the universe lifeless. Because of the unlikelihood that all these things could be just right by chance, atheists have finally had to resort to the assumption that there are countless other universes in existence in order to make it look plausible that there should be even a single one such as ours if there is no Designer.[7]

Besides this evidence of the fine tuning in the universe as a whole, Hugh Ross has recently assembled from the scientific literature a very impressive list of features for our earth, its sun, moon, and other environment that are so finely tuned as to suggest we should not expect even one planet capable of advanced life in our universe unless a Designer has put it there.[8]

In addition to the unlikelihood (if there is no Creator) of a universe or planet existing that is hospitable to life, the origin of life itself is an enormous stumbling block to atheism. Even the simplest living cells are very complex mechanisms for which a hundred million pages of instructions would scarcely suffice to provide the specifications necessary to construct one. Yet these are supposed to have arisen (by chance) very quickly and early in earth's history when the planet had barely cooled off enough so as not to cook meat. Both the complexity of living things and their sudden appearance on earth suggest the work of a Designer, not the "blind watchmaker" chance. Throughout the whole history of life, a solution to the problem of the origin of

---

[6]This history is sketched in Hugh Ross, *The Fingerprint of God*, together with evidence for a universe with a beginning.

[7]P. C. W. Davies, *The Accidental Universe* (Cambridge: Cambridge University Press, 1982); John D. Barrow and Frank J. Tipler, *The Anthropic Cosmological Principle* (Oxford: Oxford University Press, 1986); Hugh Ross, *The Creator and the Cosmos* (Colorado Springs: NavPress, 1993).

[8]Ross, *The Creator and the Cosmos*.

complex, functional information in living things seems to be far beyond the resources of a universe in which only chance and survival are at work.

Theistic evolution avoids some of these problems. With an all-powerful and supremely intelligent God overseeing everything that happens, it is not hard to imagine that he could engineer combinations of events occurring that, while not strictly miraculous, would otherwise not be expected to happen in a universe only a few billions of years old.

Theistic evolution need not do so badly in interpreting Genesis 1 either. Terms like "let the water teem" or "let the land produce" might well be understood to mean that God was providentially guiding natural processes, as we think he is when the Bible speaks of God causing the sun to rise and the rain to fall (e.g., Matt. 5:45). Even the phrase "after their kind" (KJV) does not necessarily mean that God separately created each category of plant and animal from scratch and that one kind could never evolve into another. Although the phrase has traditionally been understood to refer to plants and animals "breeding true," the word "after" is a rendering in the King James Version of the Hebrew preposition *l*, which means "according to," and the phrase in its various contexts seems to refer to classification rather than reproduction. The point of the narrative seems to be that God made the various kinds of plants and animals, without explicit comment on how he did it. All this is to say that Genesis 1 does not necessarily rule out some kind of theistic macroevolution.

The biblical problems for theistic evolution, as I see them, arise in Genesis 2. Here, many theistic evolutionists resort to claiming that these accounts are parables or allegories (fictitious history), because otherwise we have a narration that includes explicitly miraculous intervention in both the creation of Adam and of Eve. According to an evolutionary scenario in which God does not miraculously intervene, humans must have developed gradually from the apes, and thus at any time there would be a whole population of such creatures, and thus no historical Adam and Eve. So the boundary between human and ape would presumably be a fuzzy one, like the boundary between the colors red and orange. This approach introduces the concept of fictitious history into the biblical narrative, which (as I mentioned earlier) seems to me to be a serious mistake. I would like to avoid introducing fictitious history either into nature or Scripture if at all possible, for unless the data we are using is reliable, how can we possibly have any assurance that our interpretations are worthwhile?

Other theistic evolutionists, however, believe that Genesis 2 gives us a literal, historical account of the origin of the human race. They see the Bible as narrating two miraculous interventions at this point. In one, Adam is a miraculous creation (rather than a providential development), perhaps directly from the dust of the earth, perhaps indirectly from dust by remodeling an existing ape. The second intervention is the creation of Eve miraculously from Adam's side. This proposal seems much more satisfactory to me than the version of theistic evolution discussed in the last paragraph. It handles the details of Genesis 2 in a more straightforward way. It fits the biblical references to an actual fall of humans into sin, which is everywhere pictured as a real action of two individuals in history. The main problem I see with the remodeled ape version of this view is that the Genesis account indicates that "man *became* a living being" (Gen. 2:7) when God breathed into him, rather than already being alive and now acquiring humanity, as would be the case with a remodeled ape.

But there are scientific troubles with theistic evolution, too. As with atheistic evolution, it has difficulty explaining the origin of *irreducible complexity*, which is so common in living things. Michael Behe has sketched a number of these in his recent book *Darwin's Black Box*.[9] Where an organ or chemical process requires a large number of parts to be just right or the thing doesn't function at all, it doesn't look like the organ or process could have been constructed by a long sequence of small changes over many generations, since the thing would be useless until complete. Instead, it looks more like we have an entire organ or process made from scratch, or the DNA for its construction suddenly turned on, or a large number of coordinated mutations happened at just the right time and place. Though I don't want to get hung up in squabbles over terminology, I would call any of these old earth creation rather than theistic evolution.

---

[9]Michael Behe, *Darwin's Black Box* (New York: Free Press, 1996).

Besides this, the fossil record seems to have too few transitions between major biological categories to fit what I would have expected from theistic (or atheistic) evolution. Rather, new types of plants and animals regularly seem to show up without any record of close predecessors. This is especially true of the so-called "Cambrian explosion," where all the major body plans (phyla) of the animals appear in just five or ten million years (more than five hundred million years ago), with nothing comparable having happened before or since. The phenomena look more like Gordon Mills' proposal[10] that somehow God has added new information to the genes, or perhaps Robert DeHaan's suggestion[11] that new genetic programs were turned on. It is possible that most living things are descended from one or a few common ancestors, but if so, the transitions look too abrupt to be purely natural phenomena.

In addition to these gaps in the fossil record, the "shape" of the record seems to be all wrong for the various versions of evolution—both atheistic and theistic—that I am familiar with. Evolution predicts that the diversity in living things will expand from the simple, most primitive life in a cone shape rather as the limbs diverge from the trunk of an elm tree. First, the original life will diverge into various species, and these will eventually become distinct enough to be grouped into several genera. These will subsequently diverge to form families, then orders, classes, and so forth, with the basic body plans—phyla—formed last. Thus, according to evolution, the "tree of life" should be formed from the bottom up (speaking in terms of the hierarchy of categories in the biological classification system). But in fact, the phyla appear suddenly at the Cambrian explosion, and then these subsequently are subdivided into the various lower order categories, so that from the Cambrian explosion onward, the biological classification system was formed from the top downward!

These are the sorts of things that convince me that some variety of old earth creationism is preferable to atheistic evolution, theistic evolution, or young earth creationism.[12]

---

[10]Gordon C. Mills, "A Theory of Theistic Evolution as an Alternative to the Naturalistic Theory," *Perspectives on Science and Christian Faith* 47 (1995): 112–22.

[11]Robert F. DeHaan, "Paradoxes in Darwinian Theory Resolved by a Theory of Macro-Development," *Perspectives on Science and Christian Faith* 48 (1996): 154–63.

[12]For a more detailed discussion, see Hayward, *Creation and Evolution*, or Newman, "Scientific and Religious Aspects."

The Integration of Science and Theology

As far as my view of the integration of science and theology is concerned, I have a few comments to make. Initially, I like the phrase "science and theology." It is common in these discussions to talk instead of "science and the Bible," and while our concern in this book is that our theology be truly biblical, the terms "science" and "Bible" are not parallel. Science can be understood as a method, an institution, or a body of knowledge. In this it is parallel to "theology" rather than to "Bible." Science is a method or institution that investigates nature, and it is also the body of knowledge that results from this study. Theology (at least, biblical or exegetical theology) is a method or institution that investigates the Bible, and also the resultant body of knowledge. Theology studies God's *special* revelation in Scripture, while science studies God's *general* revelation in nature. If biblical Christianity is true (as I believe), then the God who cannot lie has revealed himself both in nature and in Scripture. Thus, both science and theology should provide input to an accurate view of reality, and we may expect them to overlap in many areas.

Of course, science and theology could be *defined* so they don't overlap. Perhaps science could be understood as the study of purely material things (e.g., matter, energy, etc.) and theology of purely spiritual things. Or maybe science could be thought of as the study of natural phenomena and theology of supernatural. But even if this were done, we would have to make an additional assumption to prevent overlap, namely, that there is no interaction between the physical and spiritual or between natural and supernatural—an assumption directly contradicted by the Bible and the Christian worldview. Besides this, I doubt that anyone on earth knows just what matter, spirit, or energy really are and what distinguishes them from one another. Even with the biblical accounts of miracles, it is not always easy to sort out what is miraculous intervention and what is providential oversight. Thus we should expect to see overlap.

Certainly the way we earlier defined science and theology would suggest there is overlap. Both science and theology will be interested in origins—of the world, of plants, of animals, of humans. Even of sin! Why shouldn't scientific anthropology and psychology be able to investigate whether something is dysfunctional about humanity, what it involves, and how it may have started? Of course, if science is conducted in such a way as to rule out the spiritual and supernatural, it might not be able to discover any answer that adequately handles the data.

Both science and theology will also study the continuing operation (or governance) of things in this world. And while the Bible mostly deals with God's oversight and ultimate control of nature and history, it is not totally silent about intermediate causes. Perhaps the various problems facing atheistic evolution are an indication that science should not be so tightly wed to the idea that intelligent causation of natural phenomena is out-of-bounds.

The Role of My View of Integration in the Controversy

My personal view on the integration of science and theology doesn't *have* to be peculiar to old earth creationists, and therefore it is not determinative of the position I hold. Some young earth creationists and theistic evolutionists share in this basic approach, feeling that the data of Scripture and the data of nature are fully trustworthy, and that some sort of harmonization exists between the proper interpretation of each. We just disagree on what that harmonization looks like and the relative weight to give to various scientific, theological, and philosophical considerations.

I am concerned, though, about the tendency I see among many influential theistic evolutionists to forbid the Bible to speak on scientific matters—they claim the Bible only answers the religious "who" and "why" questions, and science only the scientific "how" and "when" questions. On the contrary, it seems to me that both science and theology (or nature and the Bible) can provide input on all four of these questions, though one source may have more to say about one question and less about another. We need to be careful about making decisions on principles of interpretation that effectively rule out the consideration of significant data.

On the young earth side, I see a tendency to forbid science to have any input even on the "how" or "when" questions so far as *origins* is concerned. This usually takes the form of an objection that science is only competent to investigate hands-on, repeatable, presently occurring phenomena. But this is not so. Some sciences, indeed, concentrate on these sorts of phenomena (physics, chemistry); other branches of science (astronomy, geology, biology) are frequently historical, seeking to use surviving data to reconstruct the past. Obviously the level of certainty available to a science goes down if the phenomena it studies are outside the laboratory, unrepeatable, or occur only in the past, but it need not go to zero. All too often we Bible-believers seem to forget that similar problems exist for Bible study, too. Our knowledge of the text of the Bible depends on the surviving data of ancient manuscripts or quotations. Its interpretation depends on our modern reconstructions of the grammar and vocabulary of languages for which no one alive is a native speaker, and of cultures that have no living representatives. In spite of this, I think we are right in believing that God has arranged things so that the information we have is enough to understand the Bible adequately, though by no means exhaustively. Why should he not have also done something similar for the information we have from nature?

With these qualifications in mind, I find a real problem with the common young earth position that much of the data of modern science relative to origins is merely an appearance of age or, in the case of light from distant objects, fictitious history. Similarly, regarding many theistic evolutionists, I have a problem with their assumption that the events of Genesis 2–3 are also fictitious history. If possible, I would like to construct an integration between science and theology in the area of origins that avoids fictitious history either in nature or Scripture. This seems to me to be more consistent with the idea that nature and Scripture are both revelations from the God who cannot lie.

2. WHY IT MATTERS

The Importance of the Topic

If we Christians are right, then it is very important that people recognize the existence of the God of the Bible, that we come to realize our estrangement from him, turn back to God and cast ourselves upon him for mercy, learning to know and trust him, and to remodel our lives in conformity with his character.

But certain events in the history of science, especially since about 1800, have raised problems in the minds of many for belief in the Bible and in Christianity. These include the growing realization that the earth is more than a few thousand years old, that there have been different sorts of plants and animals on earth during its long history, and that these form a progression that has generally been accepted in scientific and secular circles as evidence that creation was the result of a natural process rather than the work of God. The three views of creation and evolution that we are discussing in this book are diverse attempts by Christians to respond to this challenge.

Historically, this particular challenge may be unique, but many other events have similarly tested faith in God. Job's experience raised real questions about the character of God. When the Babylonians destroyed the Jewish state and temple, many Israelites must have wondered what happened to the promises of God. Until Jesus' postresurrection appearances, his disciples were devastated by his death and had begun to wonder whether he really was the promised Messiah. The fact that Jesus didn't return about A.D. 1000 (or even A.D. 100) when many expected him to must have also raised doubts. Some such events have challenged particular individuals; other events have affected many at once. Some events have been so unsettling and painful that people have turned away from God as a result. But God wants us to respond to these in the right way, to learn to trust him, not giving way to despair, not becoming bitter, not refusing to face facts, not shutting ourselves up in ghettos, not bending the truth to fit with some particular group we are trying to please, but "to act justly and to love mercy and to walk humbly with [our] God" (Mic. 6:8).

Besides satisfying our own questions about the truth of the Bible and Christianity, this controversy matters because we want to help others know Jesus as their Savior and best friend. We want to reach correct and convincing conclusions that we can share with others, and so remove stumbling blocks that may be keeping seekers from coming to Christ. We also want to vindicate God's honor regarding what he has done in nature and Scripture.

Broader Cultural and Intellectual Implications

As a believer in the God of the Bible, therefore, one of my main concerns regards the influence that belief in atheistic evolution has had on our society. I see the modern tendency toward secularization in our society to be much encouraged by this belief. It has led many to ignore God, resulting in enormous distortions in public, family, and private life. It has undermined moral standards, which (in a universe with nothing more authoritative than society) can hardly have stronger sanctions than "don't get caught."[13]

In addition, it has caused many Christians to be fearful of science, of academia, and of the intellect in general, greatly weakening the outreach of the church to those who are intellectually gifted. It has led many Christians to set faith over against intellect.

Our scientific, educational, and media establishments have been heavily influenced by the idea that scientific knowledge is public fact, real knowledge, objectively true; in contrast, religious knowledge is at best subjective opinion—true for you—and at worst divisive, biased, and obstructive to real progress. It seems to me that both the young earth creationists and theistic evolutionists have unwittingly tended to reinforce this outlook—the one by dismissing large areas of scientific data, the other by effectively allegorizing the chief biblical account of origins. We old earth creationists are trying (I hope) to give a fair and straightforward reading both of the data of nature and Scripture, using standard commonsense procedures that should characterize all truth-seeking, whether mundane or academic. May the Lord help us in this!

As I see it, the young earth creation position is a stumbling block to seekers, some of whom know enough of the scientific data that they feel themselves confronted with a choice between rejecting the Bible or rejecting a fair treatment of the data.

Some theistic evolutionists believe that the Bible is without error in all that it affirms, but many do not. Since we tend to judge views other than our own by the "bad guys" in the other

[13]See the helpful discussion in D. A. Carson, *The Gagging of God: Christianity Confronts Pluralism* (Grand Rapids: Zondervan, 1996).

group, many of us see theistic evolutionists as weakening the authority of Scripture, and as providing a way out for those who are drifting away from the Christian faith. To be fair, however, we should also probably see theistic evolution as a *way in* for those who are seekers.

My Personal Journey

Regarding my own journey in relation to the creation-evolution controversy, I probably believed in some sort of young earth view when I was very young. After I had grown up, I found an old picture I had drawn back in elementary or junior high school that showed I then believed in the gap-theory, that hybrid between young earth and old earth creation. During junior high and high school I remember facing some challenges to my beliefs from evolutionary ideas, but I did not know how to respond to them.

As an undergraduate at Duke, the main challenge to my faith was the claims from liberal theology that the Bible was actually the work of superstitious or conniving people rather than those who had really observed God's intervention in history. At that time, I was helped by the writings of C. S. Lewis, the first intellectual I had encountered that was an orthodox Christian.

Later, during graduate work at Cornell, Herman Eckelmann helped me to see how an old earth model fit together the data of nature and Scripture. He also shared with me some of the solid evidences for the truth of Christianity. Through him I met Allan MacRae, who encouraged me greatly by showing that liberal critical theories about the Bible were not based on historical evidence.[14] As a result, I entered seminary to study the Bible under him.

My work both in science and theology in the years since then have amply confirmed me in the belief that the data of nature and Scripture are fully trustworthy, and that something like old earth creation is the way to go in resolving the questions of origins.

---

[14]Allan A. MacRae, *JEDP: Lectures on the Higher Criticism of the Pentateuch* (Hatfield, Pa.: Interdisciplinary Biblical Research Institute, 1994).

3. PHILOSOPHY OF SCIENCE

I am concerned about the prevalence of the idea that science by definition excludes the supernatural. I think this is an unnecessary restriction that distorts the results one can obtain from an examination of the data of nature. If any supernatural events have taken place in the history of our universe (as I believe they have, from my study of the Bible and of nature), and if they have had any profound effect on the course of events in that history (as I also believe), then the insistence that science must always assume a natural explanation for every event means that science is no longer seeking to understand what really happened. This, to put it mildly, is devastating. Scientists, by profession, should be truth-seekers. (So should theologians, and Christians in general.)

By contrast, I understand science to be an attempt to understand the phenomena of nature: (1) by examining the relevant data; (2) proposing models to explain the data; (3) testing them by comparing predictions with further data; (4) evaluating competing models; (5) seeking an inference to the best explanation; and (6) without excluding the hypothesis that God has revealed himself in this particular phenomenon. As I understand it, this is along the lines of the proposals of Stephen C. Meyer and Alvin Plantinga,[15] what the latter calls "theistic science." We are not proposing that Christians will somehow come up with a different mathematics than non-Christians, that somehow 1 + 1 + 1 = 1 (as some have misrepresented the doctrine of the Trinity). Rather, we believe that an examination of the data of nature (without assuming that God never intervenes) will lead to the solution of some long-standing problems of science that have been swept under the carpet in evolutionary discussions.

Scientific explanation should be seeking to tell it how it is. Obviously, all of us are finite beings, limited to observations made from inside the universe. Even our best perspectives may be seriously limited. But this should only serve to remind us that we need to depend on God in this endeavor as in every other.

---

[15]Stephen C. Meyer, "The Use and Abuse of Philosophy of Science: A Response to Moreland," *Perspectives on Science and Christian Faith* 46 (1994): 14–18; Alvin Plantinga, "When Faith and Reason Clash: Evolution and the Bible," *Christian Scholar's Review* 21 (1991): 8–32.

How My Philosophy of Science Informs My Approach to the Bible and Theology

The Creator of heaven and earth is also the Creator of the Bible. Thus it would seem that if both nature and the Bible are properly understood, there will be no final conflict. Admittedly, it is possible that we as finite humans might never in this life have access to enough information to see how both fit together. But the tenor of the Bible's remarks about our culpability as humans for not seeing God in nature (Ps. 19; Rom. 1) suggests that the fault lies with us, not with God or the data.

God is the maker of humanity, of our minds, our curiosity. He commends diligent study of reality, and he expects us to judge fairly; thus, it is reasonable to suppose that he is not opposed to our sincerely trying to find out how things are.

There is a strong similarity between theological method and scientific method, at least as I have experienced them in my studies both in graduate school and seminary. Both are actually (or should be) refined applications of common sense to the data of the relevant fields, as is true of academic research in any field. In both, the data should take priority over theory (the text of Scripture over personal or institutional doctrine). Of course, theorizing is necessary in both to organize and understand the data; but we need to be constantly aware that our theorizing may have misunderstood or improperly organized the data. So we need to be on the lookout for evidence of this, rather than devising interpretive principles that effectively explain away recalcitrant data. This is to say that both theological method and scientific method should be truth-seeking rather than self-affirming in one way or another.

It seems to me that the boundaries erected by the universities between the various academic disciplines are artificial. Though sometimes helpful, and probably necessary in view of the limited capacity of our minds and the vast amounts of data, they can be misleading. Some of the disciplines, including science, philosophy, theology, history, art, mathematics, and even engineering, make universalistic claims—"My discipline applies to everything!" In fact, many of these disciplines overlap very substantially and provide alternative perspectives and interacting claims about what is going on. The recent rise and fruitfulness of interdisciplinary studies suggests that we should be very careful about assuming that science and theology, nature and Scripture, are noninteracting and should stay in their own bailiwicks.

4. THEOLOGY AND SCRIPTURE

I believe that God loves the inquiring mind, and that he will reveal himself to those who really want to know the truth.

Neither nature nor Scripture is constructed to keep people from coming to the truth, but rather to lead them to it. This is why theologians have traditionally called nature "general revelation" and Scripture "special revelation."

Since both nature and Scripture are intended to reveal rather than conceal, we need to be careful to give both a fair hearing. We should not let our desire to harmonize run roughshod over the actual data of either. We should not adopt interpretive principles that allow us to explain away the data—just as any text can be allegorized, so the age of any rock can be seen as apparent age only. We should have very good reasons before we so treat any text or rock!

Having said this, we should add that both nature and Scripture may contain features designed to humble human pride. Recall Jesus' remarks in Matthew 13 regarding one of the purposes of parables. But nature as well as Scripture, as David Bossard suggests, may provide "sharp points" to goad us when we have gotten off the path.[16]

This is perhaps a good place to comment on the theological aspects of the question whether God has intervened miraculously in the events of creation. In my first section, I sketched some scientific reasons—evidence from nature—for thinking that he has.

If I understand Howard Van Till correctly, he has proposed instead that God's miraculous intervention into history is limited to his activity in redemption, and that God's mode of activity in creation is wholly providential. Van Till characterizes his view as preserving the "functional integrity" of nature, and he sees those models that involve miraculous intervention as somehow appealing to a God of the gaps.

---

[16]David C. Bossard, "Sharp Points: God's Conspiracy to Evangelize the Inquiring Mind," *IBRI Research Report* 46 (1997).

Now *providence* is the theological term for "God's oversight and guidance of events in nature and history that do not involve his overriding or superseding natural laws." *Miracle* is the corresponding term for "God's special intervention into nature or history, in which the natural laws are used in ways they would not normally function" (i.e., they are circumvented, superseded, or overridden).

Most orthodox theologians believe that God exercises providential control over all events that take place in the universe, though this somehow allows for the free moral agency of humans and various spirit beings. (We will not enter into the Calvinist-Arminian dispute here!) All orthodox theologians believe that God has intervened miraculously in history to perform miracles in the careers of Moses, Elijah, other Old Testament prophets, Jesus, and his apostles. Some of these miracles were performed through the mediation of the prophet (e.g., turning the rod into a snake; calling down fire from heaven; multiplying the loaves and fish; and raising the dead). Others took place without such mediation (e.g., the burning bush; voices from heaven; angelic appearances; light and voice on the road to Damascus). The sort of miraculous intervention of God in creation that would be envisioned by most orthodox theologians would naturally not have involved human mediation (as there were no humans around yet), though it might perhaps have involved the mediation of other spirit beings.

With regard to the question of whether God has intervened miraculously in the history of the universe, we now have a rather broad spectrum of views. At one end would be atheists who deny that God has intervened because there is no God. Next to them are liberal theologians who deny that God has intervened because their God is not the sort who would intervene. Then come the traditional deists, who admit that God intervened in creation, but deny that he intervened thereafter in history. Somewhere in this area is Van Till, who denies that God intervened in creation, but admits that he intervened in (redemptive) history. Then come many theistic evolutionists and both old earth and young earth creationists, who believe that God intervened miraculously both in creation and redemption, though they differ on the number of such interventions. Finally, we have some charismatic and Reformed theologians who see

God intervening in everything, either by greatly multiplying the miraculous or by denying the distinction between providence and miracle.

How are we to decide among these options? We have sketched in the first section of this essay some evidence from nature that points to miraculous intervention into the events of creation. The question of God's intervention in redemptive history turns on the whole matter of Christian evidences and the reliability of the Bible, which we have no space to enter into here; in any case, this is not a point of contention between Van Till, on the one hand, and Nelson and myself on the other. That the number of miracles should be seen as nearly infinite (i.e., each person can expect a miracle every day) seems to trivialize what the Bible characterizes as unusual, marvelous, powerful, and astonishing. That we should make no distinction between miracle and providence also seems to go against this same linguistic evidence.

Let us, for the moment, grant with Van Till that God has only intervened miraculously in *redemptive history*. Is it then true that he has not intervened in creation? That will depend on how human-centered we understand creation and redemption to be. The Genesis account says nothing about the creation of the angelic beings, yet Nehemiah 9:6 describes them as created. The author of Hebrews characterizes the heavenly tabernacle as not being of "this creation" (Heb. 9:11). Perhaps, then, the heavenly beings are not a part of our creation but of something earlier. And Satan has already fallen when he appears as the snake in the Garden. It may well be that the creation events of Genesis 1 take place after sin has raised its ugly head. In that case, the creation of our universe may be one of God's early acts in redeeming all of reality (including the unseen world created before the fall of Satan) from the effects of satanic rebellion. If so, the events of Genesis 1 and 2 are a part of redemptive history.

On the other hand, if we take redemption to have its usual meaning in Christian theology—God's activity to restore the human race after its fall into sin—why should we think that God will only use miracle in this realm and not in that of creation? Perhaps because we have been spooked by the problem of the God of the gaps.

A common characterization of the warfare between science and theology in Christendom has been as follows: Bible believers have typically thought that inexplicable phenomena were caused by God, but again and again science has been able to show that these events were caused by the operation of some natural law, thereby continually decreasing the gaps in which God is thought to operate and finally leaving him with nothing to do. Thus argue Andrew Dickson White, Richard Dawkins, and Stephen Hawking. Van Till is rightly concerned about this, pointing out repeatedly that God is active providentially in all phenomena. And Christians should be careful not to ascribe to God's miraculous intervention that which he has accomplished by means of natural laws.

But those who reject the testimony of the Bible often use this same sort of argument against miracles in redemptive history. Yet Van Till would want to defend these miracles and thus could equally well be accused of defending a God of the gaps.

In a sense this whole dispute is one of the God of the gaps on both the supernaturalist and naturalist sides. Those who deny the existence of God (or his activity in nature or history) postulate their God of the gaps—unknown natural law—to fill in anywhere *known* natural law will not do the trick. One of the points I was trying to make in my first section is that there are some very striking phenomena in cosmology and biology that look like the activity of a mind rather than natural law. Is Van Till consistent in trying to avoid the God of the gaps in creation while trying to defend it in redemption?

Another problem often raised against God's miraculous intervention in creation relates to style or aesthetics, and this seems to me to be reflected in Van Till's concept of the universe as having "functional integrity." For one who is responding to the idea of God as watchmaker, it seems rather gauche or unaesthetic for the watchmaker to be continually intervening in his creation, as though he didn't get it right the first time and has to be forever making adjustments.

There is something to be said for this if our picture of God as watchmaker were the whole story, but it isn't. Of course, we would think God a better watchmaker if he would design the universe to run by itself. And who wants to postulate that God, who does all things well, made a universe without "integrity"? But suppose God did not design the universe to be an automaton but rather a machine that is intended to have input. Perhaps he made the universe to be like a guitar rather than a watch, designed for performer input. In any case, Van Till's position defends God against gaucheness only for the events of creation and not against gaucheness in redemption. The proposal I am here suggesting tries to defend him in both realms with a single response.

Indeed, it may be that our universe is something like an interactive game in which the distinguishing characteristic of persons (God, angels, humans), as opposed to impersonal things (stars, rocks, trees), is that the former have the freedom to initiate actions that cannot be predicted by a knowledge of the natural laws that otherwise govern the system. In that case, our universe is like a guitar, or an interactive game, or even a TV channel in which knowledge of the behavior of electromagnetic waves cannot explain the information in the signal that the waves are carrying.

This last suggestion does indeed seem to be one of the pictures by which the Bible describes our created world (see Psalm 19). It is designed so as to carry a message about its created nature rather than to suggest that it has always existed or has merely happened by chance. And humans are morally responsible to draw this conclusion (Rom. 1).

Given the biblical teaching on the rebellion of humankind, sin has a serious effect on our willingness to handle the data of reality fairly. Yet sinfulness and finiteness will affect both science and theology, both Christians and non-Christians. And given the biblical teaching on regeneration, we who are believers should expect to find some real and growing improvement in our understanding of the things we have studied, if we will trust God with our lives and let him guide us as we grow in grace and in the knowledge of him.

5. EPISTEMOLOGY

The Role of Extrabiblical Scientific Evidence and Arguments in Natural Theology

With regard to the role of extrabiblical scientific evidence and arguments in natural theology, my studies of nature strongly confirm the biblical claim that such evidence exists. As noted above, I see this especially in cosmology, in evidence of design in both inanimate and animate nature, in the origin of life, and in the detailed history of life. This all points to the existence of an infinite, personal God.

It seems to me that many Christians in this century have made a serious mistake in denying or downplaying the significance of general revelation and natural theology, given the biblical statements on this subject and the scientific evidences that have been discovered. I haven't analyzed all the reasons for this. Part of the problem, I suspect, was the strong blow that was delivered to the natural theology of William Paley and the Bridgewater Treatises by the rise of Darwinism. The influence of Kierkegaard, Kuyper, Barth, and Cornelius Van Til in opposing natural theology has also been significant.[17] The result of this has been to keep science and Scripture apart, quite broadly so in Western society and even in evangelical and fundamental circles. Some give more weight to science, which they see as objective, while Scripture is much more fuzzy. Others emphasize Scripture as inspired, but scientists as fallible. Both approaches seem problematic to me.

The extrabiblical scientific evidence and arguments from natural theology mesh with other evidence from history, fulfilled prophecy, and transformed individual lives that confirm biblical theism. From the Bible we learn more about the existence of powerful forces of rebellion and evil that are being allowed to work out their consequences in order to teach us important lessons and the nature of God's response to all this.

Resolving Apparent Tensions Between Science and Theology

While both general and special revelation are messages from the God who cannot lie, I am reluctant to weight one as more important than the other. They are, of course, different in a number of ways.

Special revelation has the advantage in that it is already cast in human language by God who sees the whole picture and pro-

---

[17]See David P. Hoover, *The Defeasible Pumpkin: An Epiphany in a Pumpkin Patch* (Hatfield, Pa.: Interdisciplinary Biblical Research Institute, 1997), for a very readable presentation of the problems with Van Til's presuppositionalism.

vides us with needed perspective and summaries. In one of his essays C. S. Lewis made an excellent point about the difference between the teacher's analogy and the student's analogy. The teacher knows his subject thoroughly and can therefore construct a picture to help us understand it, which is profoundly correct (though it may not capture all aspects of the object it is depicting). The student does not yet know the subject thoroughly, and the pictures constructed by him may therefore contain serious errors of which he is unaware. Since God is the teacher and we are the students, we should therefore be very careful lest we dismiss the material he supplies in special revelation by assuming it is mistaken, or effectively explain it away by assigning it to a nonliteral genre without adequate warrant. For the same reason, however, we must be careful not to explain away general revelation. This is why I feel that postulating "fictitious history" in either set of data is a serious error.

General revelation, however, has the advantage of being an enormously larger body of data than special revelation; therefore, it is able to provide far more detail than Scripture. But since it is not in human language, it is more liable to misinterpretation than is special revelation.

It seems, then, that harmonization should be our ultimate strategy. Of course, we need to realize that at a given point in history, we may not yet have enough information to be able to make a proper harmonization.

In principle, I try not to bend or distort either set of information; both general revelation and special revelation are from the God who cannot lie, and thus I presume both to be reliable. Since I am an apologist as well as an interpreter, seeking to point people to the God who really exists, I am probably too prone to try for a preliminary harmonization even when the evidence is scanty.

A LETTER TO SUSAN

Dear Susan,

On the one hand, Susan, you don't really *need* to accept one position over another; God has done what he has done, and our beliefs cannot in any way change this. On the other hand, we should realize that God wants us to *do our best* to understand what he has done. An important part of our spiritual growth is an increasing understanding of God's character and works. This will be at least a lifelong project, or more likely one that will last forever!

As for what you ought to do right now, Susan, it is probably a good idea neither to hurry into any one of these positions nor to reject them all. The dangers of dogmatic insistence on one of them or agnostic refusal to adopt any of them seem to me to be equally great, sort of like the leaven of the Pharisees versus the leaven of the Sadducees (Matt. 16:5–12). We need to have some appreciation not only for the content of our beliefs and why we believe them, but also for the degree of certainty that is warranted for each of them. In this way, it will be easier for us to see what we need to change if it becomes apparent that we are wrong in some area.

For any one of these three positions we're discussing, I think that if we already embrace the position, it's OK to continue to do so while we look around. As we look around, we want to get some feel for what the main questions are, what the data looks like, what the advantages and problems of each view are, what principles for deciding such questions various people are using, and particularly whether we personally are using a double standard, that is, whether we are applying tougher tests to other views than we are to our own. The Lord wants us to be faithful to him and fair to our opponents. I hope this book will provide some help for each of us as we look around.

As created beings with limited capabilities, dependent on what God has revealed both in nature and Scripture, we shouldn't be surprised if we don't understand how everything fits together. That's OK. God isn't asking us to run the universe while he takes a nap! Surely he wants us to get squared away on the most important and basic questions first: Is there a God? What is he like? What does he expect from me? The *Good News* is that the God who actually exists is the God of the Bible; that he loves us (though we may not yet care much for him); and that he has provided a way to return to him and know him as our best friend through the work of Jesus, who takes the punishment we deserve and provides the righteousness we ought to have. If we've got this right and have made the right response to God's offer, then the main questions are settled and we have breathing room to spend the rest of our lives on the lesser questions, like how we ought to live, the mode of baptism, the form of church government, the time of the Millennium, and the nature of God's work in creation.

As I sketched above, I do think that we have excellent reasons for believing that God exists, that he is the God who wrote the Bible, and that several of the major dilemmas facing science today are resolved when this is recognized. I have no qualms about using the Bible to understand the universe, human history, my own significance in the scheme of things, and how God is going to bring this current phase of history to its conclusion.

Well, Susan, that's about it. You should accept my view if it seems to you to be the right one. You shouldn't be surprised that good Christians disagree. Look at all the things good Christians disagree about! Nor should you throw up your hands in despair of ever answering questions that cause disagreement among good Christians with more education than you. We will need to answer to God for *our* actions and beliefs, not for someone else's. Every one of us can think of times when we have believed things, made decisions, or taken actions for reasons that later turned out to be rather poor ones. What does the Lord require of us but to act justly and to love mercy and to walk humbly with him (Mic. 6:8). May he help each of us to do so!

RESPONSE TO ROBERT C. NEWMAN

Walter L. Bradley

I would like to begin by indicating that I agreed with essentially everything presented by Dr. Newman. Therefore, I will use my critique to expand on Newman's discussion of the God-of-the-gaps issue and evaluate Howard Van Till's argument that a creation with so-called functional integrity is somehow superior. This presentation would be best read after reading my critique of Van Till's chapter.

I would like to give a brief picture of the design features of our universe from the point of view of a mechanical engineer and then consider whether this design would be more impressive if it were the result of a "gap-free" process, which Van Till calls a creation with functional integrity. God made the universe to be very dynamic, interesting, and unpredictable by including in the laws of nature some physical processes that are linear and other processes that are nonlinear. The net result is that we have some phenomena, like the earth's movement about the sun, which are the result of linear processes that are very stable and periodic, and other physical processes, such as weather cycles, that are the result of nonlinear processes and therefore are very unstable, leading to unpredictable behavior. A universe of all linear processes would be extremely static and boring while a universe of all nonlinear processes would be too chaotic to make a suitable home. Our universe is carefully balanced between stability processes that make life generally predictable, but with dynamic features that give persistent changes, surprises, and progress.

In such a universe, living systems need to be robust to be able to adapt to the constantly changing environment. I believe that God incorporated this capacity for robustness in living systems to match the continuously changing environment by including genetic diversity in living systems and by allowing further modification of this diversity through mutations. Thus, I believe that microevolution, which I mean to include both changes in genetic population distributions within species as well as mutations that modify existing characteristics (as distinct from the more dubious claim of mutations that create entirely new characteristics), are part of God's systemic design. Such a design provides dynamic, adaptive change in response to a continuously changing biosphere, allowing for an endless variety of interesting developments in God's creative story.

But what about Van Till's claim that creation is somehow more gifted if all events are done with creation on "autopilot," all developments somehow being incorporated into the initial design scheme? I believe such a design would substantially compromise the universe that we have in one of several ways. First, if the provision of information is to come through the assignment of properties to matter, then the outcomes that are possible will be significantly limited by the initial property assignments given. If, for example, the sequencing of amino acids in proteins were due to the chemical bonding preferences, then only one or a few sequences would be permissible, severely limiting the varieties of proteins that could be produced. On the other hand, when this information is provided by other means, the number of ways that biopolymers (such as proteins) or systems (such as living cells) can be organized is indeed unlimited.

Second, the constraint of trying to put all of the information into the initial properties may have some very significant performance penalties that are not apparent at first glance. Suppose that I wanted to design an automobile that could self-assemble. It would certainly be possible in principle to make such an automobile. However, the degree of complexity associated with these additional requirements would greatly increase the cost and would almost certainly compromise the performance, since these additional capabilities come at a high cost of additional complexity that is useful only in the assembly but not thereafter. In the same way, there may be some significant design compromises in a universe that is able to unfold with all the necessary information incorporated into the properties of matter.

In summary, there is no rational basis for Van Till's claim that a universe that unfolds entirely on autopilot represents a better design or a more fully gifted creation by God than one in which not all of the necessary information is imparted in the properties of matter alone but is incorporated at certain critical points in the developmental history of the universe.

RESPONSE TO ROBERT C. NEWMAN

John Jefferson Davis

Of the three contributors to this symposium, my own position is closest to that of Robert Newman. Like Professor Newman, I feel comfortable with the term "progressive creationism," which understands God's creative activities to have occurred in a number of steps over long periods of time.[1]

I also agree that some form of a complementarity model is helpful in understanding the overall relationship between science and religion.[2] Such a perspective recognizes that Scripture and science have different purposes and foci, with Scripture focusing on qualitative issues of purpose and meaning, and the sciences concentrating on quantitative issues of process and structure. Both Newman and I would recognize that because God is the Creator of nature, and the author both of nature and Scripture, some areas of overlap between scientific and biblical concerns are to be expected. We both expect that God's works of creation, providence, and redemption may in certain instances leave recognizable traces in the natural order since they take place in the real world of space and time, not in some purely private realm of human subjectivity.

---

[1]I would, however, tend to see more evidence of transitional forms in the fossil record than does Newman, as will be indicated in the response to follow.

[2]For discussion of the complementarity model and other models of relating science and religion, see Ian Barbour, *Religion in an Age of Science* (San Francisco: HarperCollins, 1990), 3–30.

Newman and I also agree in our opposition to methodological naturalism as a philosophy of science.[3] That is, in our view, there is no a priori reason why supernatural factors should be excluded by definition in attempts to understand the features of the natural order. We believe, for example, that divine activity is the "best explanation" of the origin of the first living cells and of the major life-forms in the Cambrian explosion, and that purely natural processes cannot provide convincing explanations of such quantum leaps in the level of observed complexity.

Both Newman and I agree that there is overwhelming scientific evidence that the universe and earth are very old. This evidence is readily available in the works of Davis Young, Alan Hayward, Brent Dalrymple, and Dan Wonderly.[4]

Unlike Newman, I understand the "days" of Genesis 1 in terms of the *literary framework* hypothesis, in which the days are not twenty-four-hour periods of time, but a literary framework for God's creative work.[5] The ordering of the days is understood to be topical and theological rather than chronological in nature.

According to Newman, the main point of Genesis 1 seems to be that God made the various kinds of plants and animals, with little explicit detail on the processes involved. I agree with Newman in his view that Genesis 1 does not necessarily rule out some possible forms of theistic evolution. I tend to see, however, more possible evidence than he does in the fossil record of tran-

---

[3]For an insightful criticism of the limitations of methodological naturalism, see Alvin Plantinga, "Methodological Naturalism?" *Perspectives on Science and Christian Faith* 49 (September 1997):143–54.

[4]Davis A. Young, *Christianity and the Age of the Earth* (Grand Rapids: Zondervan, 1982); Alan Hayward, *Creation and Evolution: Rethinking the Evidence from Science and the Bible* (Minneapolis: Bethany House, 1985); G. Brent Dalrymple, *The Age of the Earth* (Stanford, Calif.: Stanford University Press, 1991); Dan Wonderly, *God's Time Records in Ancient Sediment* (Flint, Mich.: Crystal Press, 1977); and *Neglect of Geological Data* (Hatfield, Pa.: Interdisciplinary Biblical Research Institute, 1987).

[5]For this view, see Henri Blocher, *In the Beginning* (Downers Grove, Ill.: InterVarsity Press, 1984), 39–59, and M. G. Kline, "Because It Had Not Rained," *Westminster Theological Journal* 20 (1957–58): 146–57. A form of this view is also found in Thomas Aquinas, *Summa Theologica*, Pt.1, qq. 65–74, "Treatise on the Work of the Six Days." David A. Sterchi, "Does Genesis 1 Provide a Chronological Sequence?" *Journal of the Evangelical Theological Society* 39 (December 1996): 529–36 argues that the use of the definite article in Genesis 1 does not require a chronological ordering of the days; the days are apparently numbered on the basis of content, not order in time.

sitions between the major forms. For example, the primitive extinct amphibian *Ichthyostega* has anatomical features that are intermediate between the more advanced amphibians and the lung fishes from which they may have been derived. New findings related to *Panderichthys*, an extinct lobe-finned fish, show a mosaic of fishlike and amphibianlike characteristics.[6]

Newman does not discuss the extensive evidence in the fossil record for ancient, extinct animals that display anatomical characteristics that are intermediate between the reptiles and the mammals. This fossil evidence is extensive, extends over a period of over 150 million years, and points to transitions from "small, cold-blooded scaly reptiles to tiny, warm-blooded furry mammals."[7]

Most paleontologists believe that the ancestry of modern mammals is to be found among a group of extinct mammal-like reptiles knows as the cynodonts. The fossil record of mammal-like reptiles is the most complete of any group of terrestrial vertebrates with the exception of the mammals themselves.

For example, one might consider the case of *Cynognathus*, an extinct animal about the size of a large dog, displaying a blend of reptilian and mammalian characteristics. *Cynognathus* had a large skull that was doglike in appearance. Its teeth were differentiated and specialized, unlike the undifferentiated teeth of a reptile. The vertebral column was differentiated into cervical, dorsal, and lumbar regions, like that of a mammal. These and other skeletal features show that it was an active, carnivorous reptile that was approaching a mammalian stage of development in many respects. *Cynognathus* was only one of many extinct tetrapods that display characteristics intermediate between the reptiles and the mammals.[8]

It is also worth observing that there are *living* animals—the monotremes, or egg-laying mammals of Australia—that exhibit characteristics intermediate between reptiles and the more

---

[6]Edwin H. Colbert and Michael Morales, *Evolution of the Vertebrates: A History of the Backboned Animals Through Time*, 4th ed. (New York: John Wiley & Sons, 1991), 67–69; Per E. Ahlberg, et al., "Rapid Braincase Evolution Between *Panderichthys* and the Earliest Tetrapods," *Nature*, 2 May 1996, 61–63.

[7]Robert L. Carroll, *Vertebrate Paleontology and Evolution* (New York: W. H. Freeman, 1988), 361. See also T. S. Kemp, *Mammal-Like Reptiles and the Origin of Mammals* (London: Academic Press, 1982).

[8]Colbert and Morales, *Evolution*, 123–25.

advanced mammals. The echidna and the platypus, or duckbill, are primitive mammals that reproduce by laying eggs (like reptiles) and yet like mammals suckle their young on milk produced by modified sweat glands. As in reptiles, the cervical ribs of these animals are not fused, and certain reptilian features can be seen in the structure of their skulls. Their urogenital systems and rectums open into a common cloaca as in reptiles, rather than separately as in mammals. In these respects the echidna and platypus represent living forms that are intermediate between extinct mammal-like reptiles and the higher mammals.[9]

Another well-known example of an intermediate form is *Archaeopteryx*, generally regarded as the oldest known bird. First discovered in the 1860s in Germany, *Archaeopteryx* has teeth and a long bony tail, as do reptiles, and yet has feathers and a fused clavicle ("wishbone") as do birds.[10] It is generally regarded as a good example of a "missing link" between the dinosaurs and and the birds. In 1997, two Argentinian scientists announced the discovery of *Unenlagia comahuensis*, an extinct dinosaur that displayed characteristics intermediate between *Archaeopteryx* and the group of dinosaurs regarded by many as being most closely related to birds.[11] Such evidences tend to give added plausibility to the hypothesis that the origins of birds are to be found among the dinosaurs.

Newman points to some of the serious theological questions that arise with respect to theistic evolutionary scenarios of human origins. His discussion would have been strengthened, however, by a more detailed discussion of the hominid fossil evidence.[12] The fossil record of the *Australopithecines*, *Homo erectus*, the Neanderthals, and other extinct hominid forms indicates that modern *Homo sapiens* did not appear on earth without precursors. These humanlike precursors were characterized by the

---

[9]Ibid., 241.

[10]Michael J. Benton, *Vertebrate Palaeontology* (London: Unwin Hyman, 1990), 209–13.

[11]Fernando E. Novas and Pablo F. Puerta, "New Evidence Concerning Avian Origins from the Late Cretaceous of Patagonia," *Nature*, 22 May 1997, 390–92.

[12]For example, see Richard G. Klein, *The Human Career: Human Biological and Cultural Origins* (Chicago: University of Chicago Press, 1989), and David Lambert, *The Field Guide to Early Man* (New York: Facts on File, 1987), for accessible presentations of the hominid fossil evidence.

transition to bipedalism, or upright walking; gradual increases in brain size; changes in skeletal and dental structures; and growing sophistication in stone tool technologies—changes that extended over millions of years. These data need to be fully integrated into any attempt to relate the Genesis account of human origins to the hominid fossil record.

RESPONSE TO ROBERT C. NEWMAN

J. P. Moreland

My own views about the creation-evolution controversy are divided between old and young earth creationism. While I lean heavily toward old earth views, I do not see the issue as cut-and-dried. My greater concerns, however, involve my desire to promote theistic design and some form of special creationism over against theistic evolution, which I take to be biblically inadequate and less than required by the relevant scientific considerations. Because of this, even though there may be a point in Newman's essay here and there about which I am not entirely persuaded, I am in substantial agreement with most of what he claims. In my response, then, I want to offer some reflections about an objection likely to be raised against his approach. Put simply, the objection says that when we explain some natural phenomenon by saying that God did it, then this "explanation" simply stops the search for further natural explanations and is not very informative.

To elaborate a bit, this objection claims that theistic science, of which Newman's progressive creationism is a specific version, is not a fruitful research program for solving empirical problems or for guiding new research and yielding new empirically testable constructs in other areas of investigation. Because of this, theistic science, while still science in principle, nevertheless, should be abandoned. Explanations in terms of divine action stifle scientific investigation, hamper progress, and short-circuit the search for naturalistic explanations and the conceptual gains that come from such a search. What shall we say about this problem?

For a moment let us grant the unambiguous importance of fruitfulness as a criterion for evaluating the merits of a scientific theory or research program (roughly, a family of theories). I have two things to say on such an assumption. First, if we grant that theistic science (e.g., progressive creationist practices and theories) have not been fruitful, I think the same thing can be said with equal force against evolutionary theory. How has evolutionary theory led to fruitful research, useful explanations, or empirically successful predictions? Various Christian and non-Christian scholars have claimed that evolutionary theory is in a period of crisis precisely because it is a dead-end research program if judged by its fruitfulness.[1]

Moreover, even if we grant that a theistic scientist's utilization of theological concepts has not fruitfully suggested new lines of empirical research (and this need not be granted), all that follows from this is that theistic scientists need to do more work developing the infrastructure of their models, not that their models are not part of natural science or that they cannot be empirically fruitful where appropriate. Now some of this development is already taking place as excitement mounts about new efforts in intelligent design research.

So far I have assumed that fruitfulness is a clear-cut, appropriate criterion for evaluating the merits of a scientific theory (or research program). Unfortunately, this assumption is far from uncontroversial. An appeal to empirical fruitfulness to judge the relative merits of young or old earth creationism versus theistic or naturalistic evolution may be question-begging and certainly represents a naive understanding of the intricacies of such fruitfulness as a criterion for assessing the relative merits of rival research programs, especially these rivals.

For one thing, two rivals may solve a problem differently depending on the way each theory depicts the phenomenon to be solved. Copernicus solved the motion of the planets by placing the sun in the center of the universe. Ptolemy solved that motion by a complicated set of orbitals with smaller orbitals (epicycles)

[1]See Michael Denton, *Evolution: A Theory in Crisis* (London: Adler & Adler, 1986); J. P. Moreland, ed., *The Creation Hypothesis* (Downers Grove, Ill.: InterVarsity Press, 1994).

contained within larger ones. Each solution was different (and not necessarily of equal effectiveness). Thus, the standards for measuring one research program may differ substantially from those relevant to its rival. I am not saying that rivals are incommensurable (i.e., that they cannot even be compared with each other). I am simply pointing out that it is often more complicated to compare rivals than is usually thought to be the case. If one rival is the dominant paradigm, the less acceptable research program (e.g., theistic science) can easily be judged a failure by standards set by its rival. And this can be Procrustean.

For example, it is possible for two rivals to rank the relative merits of epistemic values in different ways or even give the same value a different meaning or application. An epistemic value is a feature of a scientific theory that, if present, increases the rational acceptability of that theory. A number of important epistemic values have been identified for scientific theories: theories should be simple (especially compared to its rival), observationally accurate, predictively successful, fruitful, have a wide scope of application, solve internal conceptual problems (i.e., contain theoretical concepts that are clear, not vague or ambiguous; plausible, not improbable or implausible; and at home in the rest of the theory, instead of odd and ad hoc in appearance), and solve external conceptual problems.

An external conceptual problem is an intellectual difficulty for a scientific theory that has its origin in a field outside science, like theoretical mathematics, history, ethics, philosophy, or theology, and which tends to disconfirm the scientific theory in question. For example, if there are ethical arguments for the fact that living things have essences or natures, then if a scientific theory most naturally implies that they have no such natures, these ethical arguments amount to external conceptual problems that must eventually be solved. Again, if there are philosophical arguments for the fact that the past could not have been infinite in duration, these present external conceptual problems for cosmological models that imply an infinite past. Thus, rivals can differ radically about the nature, application, and relative importance of a particular epistemic value.

Creationists and evolutionists do not need to attempt to solve a problem (e.g., a gap in the fossil record) in precisely the same way, nor do they need to employ the same types of solutions or the same ranking of epistemic values in their solutions. Creationists may elevate the value "solve theological or philosophical internal and external conceptual problems" above the value "offer solutions yielding empirically fruitful lines of new research." For example, creationists may feel they have good intellectual grounds from philosophy, theology, and biblical exegesis for thinking that the general theory of evolution (theistic or naturalistic) is wrong. They may think it more intellectually responsible to find a model of origins that solves these external conceptual problems than to develop views that are empirically fruitful. There is nothing unscientific about this at all, and it is question-begging to claim that a criterion of empirical fruitfulness set by one research program (e.g., the search for evolutionary mechanisms) should be most important for a rival program and, if not, the rival is not even science or, if science, is not as rationally acceptable as its (allegedly) more empirically fruitful competitor.

Theistic science may be more apt in solving conceptual problems than a rival research program, and this may be more important to theistic science than empirical fruitfulness. More work needs to be done here, but even if I am partially correct, then given the current preference for the empirical over the conceptual, and given the general lack of appreciation of the role of conceptual problems in the history of science, theistic science is likely to be given a negative epistemic assessment beyond what it may deserve.

Finally, one rival will sometimes consider a phenomenon basic and not in need of a solution, empirical or otherwise. It may, therefore, disallow questions about how or why that phenomenon occurs and, thus, can hardly be faulted for not being fruitful in suggesting lines of empirical research for mechanisms whose existence is not postulated by the theory. As Nicholas Rescher has pointed out,

> One way in which a body of knowledge S can deal with a question is, of course, by *answering* it. Yet another, importantly different, way in which S can deal with a question is by disallowing it. S *disallows* [Q] when there is some presupposition of Q that S does not countenance: given S, we are simply not in a position to raise Q.[2]

[2]Nicholas Rescher, *The Limits of Science* (Berkeley, Calif.: University of California Press, 1984), 22.

For example, motion was not natural in Aristotle's picture of the universe and, thus, examples of motion posed problems in need of explanation. But in Newton's picture of the universe, uniform linear motion is natural and only changes in motion pose problems in need of solution. Thus, suppose a Newtonian and an Aristotelian are trying to solve the observational problem of how and why a particular body is moving in uniform linear motion. The Aristotelian must tell how or why the body is moving to solve the problem. But the Newtonian can disallow the need for a solution by labeling the phenomenon as a basic given for which no solution utilizing a more basic mechanism in response to a how-question is possible.

Similarly, certain phenomena, like the origin of life or gaps in the fossil record, are not problems in need of solution for creationism beyond an appeal to the primary causal agency of God. But they are problems for evolutionary theory, and fruitful lines of research for new mechanisms must be sought. However, it is naive and question-begging to fault creationists for not developing fruitful problem-solving strategies for such gaps compared to their evolutionary rivals because such strategies are simply disallowed given that these phenomena are basic for creationists. In this case, it is enough for creationists to use theological notions to guide them in the search for scientific tests to establish the phenomena predicted by the theological constructs. Once the what-question is answered, there is no further material or mechanistic how-question that arises.[3]

Finally, in certain cases advocates of theistic science can be more open than their methodological naturalist counterparts to allowing the empirical data to speak for themselves. This is espe-

---

[3]It should be added that some appeals to fruitfulness actually distort an intellectual issue. For example, William Bechtel argues that of two views regarding the relationship between mental states and brain states—the two are different but correlated (property or substance dualism) or the two are identical (type or token identity physicalism)—the identity thesis can be judged as the superior position based on the fruitfulness of the scientific research program that follows from it. See his *Philosophy of Mind* (Hillsdale, N.J.: Lawrence Erlbaum Associates, 1988), 101–3. However, this recommendation distorts the proper order of analysis regarding the mind-body problem (philosophy is more basic and important than is science), and in any case, whatever research program the identity thesis generates, the same research program, perhaps with very minor modifications, could be generated from the dualist correlation position.

cially true in areas where God could have acted through primary or secondary causation, even if, theologically speaking, the former is more likely. As an example of this, consider the research by Kok, Taylor, and Bradley.[4] In their view, the type of functional specificity in the amino acid sequences in the proteins of living organisms is due to the primary causal activity of a designer. Based on this assumption, they predicted and "verified" the fact that different forms of self-ordering tendencies in matter, especially steric interference that allegedly gives rise to nonrandom preferential amino-acid sequencing in proteins, are absent. If such interference had been found, they would have seen God's activities in designing these sequences as secondary causes. However, the advocate of methodological naturalism (e.g., naturalistic or theistic evolutionists) would seem to be in a position of *requiring* such steric interference, or something of this sort, due to his or her commitment to methodological naturalism and secondary causation.

Advocates of theistic science like Newman do not believe in a capricious God nor do they employ willy-nilly the notion of a miraculous creative act of God. In their view, such acts are in the minority, and allowing for their possibility (and appealing to them to guide research, formulate predictions and expectations of what we will discover in the world, and explain certain facts) in no way hampers the general search for natural mechanisms. And the claim that creationist theories are not fruitful cannot be employed as a conversation stopper to silence those like Newman who are content to give their arguments, appeal to the evidence, and let the chips fall where they may. What old and young earth creationists are trying to do is open people's minds and show that there is a legitimate issue here: When the evidence is actually examined, are naturalistic mechanisms adequate to explain the living world or are the intentional actions of an intelligent designer a better explanation? After all, the real issue in this debate is not fruitfulness but truth, and when it comes to macroevolution, young and old earth creationists are agreed that the best evidence available justifies the claim that intelligent design explanations are not complementary to evolutionary mechanisms, and the former are true while the latter are false.

---

[4]R. A. Kok, J. A. Taylor, and W. L. Bradley, "A Statistical Examination of Self-Ordering of Amino Acids in Proteins," *Origins of Life and Evolution of the Biosphere* 18 (1988): 135–42.

RESPONSE TO ROBERT C. NEWMAN

Vern S. Poythress

Dr. Newman's general approach as an old earth creationist is attractive in that he tries to exercise care in interpreting both the Bible and modern science. In particular, Newman's general remarks about the relation of science to theology are sound and helpful.

Newman rightly sees that *for the most part* the Bible does not go into detail about the means that God may have used in his acts of creation. Since the Bible supplies only limited information in this area, some kinds of theistic evolution are theologically permissible. Newman rightly bases his preference for special creation primarily on scientific data.

Newman is also right in warning against dissolving biblical claims. He points to the exceptional character of the creation of Adam and Eve. As he observes, the idea of "fictitious history" or "allegory" will not work for Genesis 2. Genesis by its overall structure gives every indication of talking about actual events. The later chapters of Genesis contain the stories of Abraham, Isaac, and Jacob, whom the Bible everywhere represents as real people. Similarly, Adam and Eve and the events involving them are real.

Unfortunately, in matters of detail Newman is not always the best possible representative for old earth creationism. In my judgment, a moderate concordist view, which I briefly discuss in my response to Nelson and Reynolds, is the best interpretation of Genesis 1.[1] By adopting a modified "intermittent day"

[1]See Derek Kidner, *Genesis* (Downers Grove, Ill.: InterVarsity Press, 1967), 54–58.

view of Genesis 1, Newman has unnecessarily entangled himself in a number of interpretive mistakes.

First, Newman claims that the seventh day of Genesis 2:2–3 is still future. But clearly it is not. The Hebrew tenses in 2:2–3 are the normal ones used to continue a narrative of *past* events, just as in 1:3–2:1. Likewise, Exodus 20:11 says that God rested (in the past).[2]

Second, Newman claims that the creative actions in Genesis 1 *began* on particular days, but continue into the present. (Some of the details are found in Newman's book, not in the present essay.)[3] Of course, Genesis 1 includes creative *pronouncements* that retain their validity up to the present. The sun and moon continue to mark the seasons, in accord with Genesis 1:14. Animals continue to reproduce, in accord with Genesis 1:24. But the creative works of making light, making the sun, and making the kinds of animals are complete: "Thus the heavens and the earth were completed in all their vast array" (Gen. 2:1). Israelite readers would so understand it, and indeed it is confirmed by ordinary observation: the things God made are already here and are fully formed and functioning.[4]

Third, Newman claims that each day is a twenty-four-hour day that marks only the beginning of a creative period: God's works of creation do not take place *during* the day, but during a long period of time inaugurated by the day.[5] This view makes no sense to an Israelite. An Israelite worked six days and rested on the seventh because he was following the pattern of God,

---

[2]Newman's book appeals to Hebrews 4 and John 5:17 (Robert C. Newman and Herman J. Eckelmann, Jr., *Genesis One and the Origin of the Earth* [Downers Grove, Ill.: InterVarsity Press, 1977], 65). But this appeal only confuses the issue about what Israelites would have understood *before* Hebrews 4 and John 5:17 had been written. In fact, the New Testament passages assume that we already know what Genesis 2:2–3 asserts about the past. Hebrews 4 indicates that God's people will enter his rest in the future, but assumes that we know that God has already entered the rest himself (Heb. 4:10b). John 5:17 tells us that God is still working his works of providence and *redemption;* we know from Genesis that he has finished his work of *creation.*

[3]Note his statement about the sixth day in his present essay. See also Newman and Eckelmann, *Genesis One,* 83–88.

[4]When Newman claims that God's creative work continues, he imposes on Genesis 1 a technical scientific meaning, instead of seeing it as an ordinary description to Israelites (Newman and Eckelmann, *Genesis One,* 84–85). See my discussion of biblical language in my response to Nelson and Reynolds.

[5]Newman and Eckelmann, *Genesis One,* 64–88.

who did it first (Ex. 20:11). By analogy with human work, Genesis 1:6–8 describes the work that God accomplishes *during* the second day, not work during a long era, at the end of which a second twenty-four-hour day finally begins in verse 8b. Likewise with the other days. So the six days, taken together, cover the entire creative activity of God (note Ex. 20:11 again). The days are not tiny twenty-four-hour periods sprinkled into a vastly larger period of creative activity. Rather, they form a structural whole, a week, during which the works of creation take place and at the end of which God rests.

Fourth, Genesis 1:16 says that God "made" the lights, not that he caused to appear already-existing heavenly bodies, as Newman states.

Newman admits that some degree of logical and topical organization occurs in Genesis 1 so that not everything is purely chronological.[6] If so, he is really quite close to a moderate concordist view, which would relieve him of his difficulties.[7]

We should also note a problem with Newman's discussion of miraculous intervention. Newman fails to slam the door hard on the idea of self-contained "nature," and thereby weakens his case vis-à-vis Howard Van Till and deistic ideas. Let me explain. Newman describes special acts of creation as "miraculous intervention." He then distinguishes miracle from providence by means of "natural law." But natural law is an exceedingly problematic term. It could be shorthand for God's law or decrees for creatures. In this sense, everything conforms to this natural law. This meaning of natural law does not succeed in distinguishing one kind of action (miracle) from another (providence). Or the term natural law could be shorthand for "what nature does when left to itself." In that case, it suggests deism, rather than biblical truth. In the final analysis, a created thing does not do anything "by itself," but only by the sustaining power of God (Heb. 1:3). Or does natural law denote a kind of "cosmic mechanism," created by God, that runs the world when God is on vacation? The Bible allows no such mechanism but declares repeatedly that God is in constant personal control through the exercise of "his powerful word" (Heb. 1:3; see Pss. 147; 121). Or

[6]Ibid., 79.

[7]See my discussion of a moderate concordist view in my response to Nelson and Reynolds.

does natural law mean, "what current scientists think the laws are"? Then it changes with the progress of science and is only an educated guess approximating God's actual commands. None of these approaches is of any help in producing a solid biblical understanding of miracle.

One could still distinguish miracles as more extraordinary, striking, and wonderful acts, uniquely providing special revelation, or perhaps as acts more central to accomplishing God's redemption. But this route will not supply Newman with what he wants.

Newman would be better off talking about "normal" and "exceptional" acts of God. The word "intervention," by contrast, has serious liabilities. It tends to suggest that God intervenes in a mostly self-sufficient mechanism. Most of the time God does little but watch. But when things come to a particular point, he "intervenes." Or else it may suggest that God made the universe in a certain way, and then "intervenes" in a manner contrary to the innate character of the created world. Both of these faulty views conflict deeply with biblical teachings about God's constant personal involvement in his world, the harmony in his plan, and the coherence among all his actions (Pss. 104; 135; 136; 147; 19; Isa. 46:9–10; Eph. 1:10–11; Col. 1:15–20).

The following more minor problems remain.

First, the alleged "biblical hints" of an old earth are quite weak and should be dropped from the argument.

Second, in evaluating the different flood theories, it is important to decide whether Noah's flood was local or global. That is, did the waters of the Flood cover only one region or the entire globe? A local flood is much easier to harmonize with an old earth position. Hence, one could wish that Newman provided further support for the idea that Noah's flood was local.[8]

Third, what about "natural theology"? Newman rightly notes that the Bible includes the idea of general revelation. Everything that God made testifies about him and leaves people without excuse for turning away from him (Rom. 1:18–32; Ps. 19). But then Newman moves too quickly to the problematic idea of natural theology. Perhaps Newman uses natural theology almost as a synonym for general revelation, but such is not its usual meaning. It can denote what a person, biased by sin,

[8]For more on Noah's flood, see my response to Nelson and Reynolds.

and perhaps refusing to submit to the Bible, can be expected to accept from general revelation. Historically, natural theology in this sense has regularly underestimated the radical effects of sin in bringing darkness and perversity to the minds of human beings (Eph. 4:17–19).

CONCLUSION

Robert C. Newman

In regard to the feedback of my colleagues, I will respond to a number of their points topically.

FOSSIL RECORD TRANSITIONS

I am aware that some scientists claim to have found some transitions between higher categories in the biological classification system. That is why I worded my comments on this as cautiously as I did. I cannot make any independent judgments on the validity of these claims because I don't have the enormous amount of specialized training and field experience in biology and geology that would be necessary for this. Even specialists frequently argue about these matters.

In any case, for my model of the creation events, there is no need to rule out all such transitions. The basic kinds of living things mentioned in Genesis may very well have been made by God from previously existing kinds rather than from scratch. That is just the sort of thing we need to test by means of scientific investigation. But it needs to be scientific investigation that is open to the possibility that some of these events may not be the result of purely naturalistic causes. Explanations suggesting that these events may not be the result of purely naturalistic causes are often treated as being out-of-bounds.

My comments on this topic were seeking to incorporate the insights of Stephen Gould,[1] David Raup,[2] Steven Stanley,[3] and others who have noted a great rarity of transitional fossils. This is hardly what one would expect if these transitions were the result of random walks of hundreds of undirected mutations in large populations so as to cross the gaps in biological diversity. This should leave the fossil record littered with a multitude of transitional fossils—which we don't find. As I suggested earlier, it looks as though, given that these purported fossils really are transitional, we are seeing either a programmed transition built into the DNA of the creature itself[4] or somehow information is being added to the DNA from outside[5] by some process far more efficient than random events. All this sounds more like old earth creation than evolution, even of the theistic sort.

Regarding anthropoid fossils that predate modern human beings, I am inclined to see these as not ancestral to us, though I am not sure where to put the break. One Christian, Glenn Morton,[6] puts the origin of the human race back several million years ago with the Australopithicines; at the other extreme is Dick Fischer,[7] who places the origin of humans about five thousand years ago with Adam, but not as the progenitor of the entire

---

[1]Stephen Jay Gould has said, "The extreme rarity of transitional forms in the fossil record persists as the trade secret of paleontology" (*Natural History* 86 [1977]: 14).

[2]David Raup has observed, "We are now about 120 years after Darwin. . . . Ironically, we have even fewer examples of evolutionary transitions than we had in Darwin's time. By this I mean that some of the classic cases . . . have had to be discarded or modified. . . ." ("Conflicts Between Darwinism and Paleontology," *Field Museum Bulletin* 30 [1979]: 25).

[3]Steven M. Stanley said that "despite the detailed study of the Pleistocene mammals of Europe, not a single valid example is known of phyletic (gradual) transition from one genus to another" (*Macroevolution: Pattern and Process* [San Francisco: W. H. Freeman, 1979], 82).

[4]See Robert F. DeHaan, "Paradoxes in Darwinian Theory Resolved by a Theory of Macro-Development," *Perspectives on Science and Christian Faith* 48 (1996): 154–63.

[5]See Gordon C. Mills, "A Theory of Theistic Evolution as an Alternative to the Naturalistic Theory," *Perspectives on Science and Christian Faith* 47 (1995): 112–22.

[6]Glenn R. Morton, *Foundation, Fall and Flood: A Harmonization of Genesis and Science* (Dallas: n.p., 1995), 238–77, esp. 249; *Adam, Apes and Anthropology: Finding the Soul of Fossil Man* (Dallas: DMD Publishing Co., 1997).

[7]Dick Fischer, "In Search of the Historical Adam: Part I," *Perspectives on Science and Christian Faith* 45 (1993): 241–50.

human race. I am closer to Hugh Ross,[8] who sees the creation of Adam as some tens of thousands of years ago, which seems to fit the evidence from mitochondrial Eve and Y-chromosome Adam better than Morton's view, and the biblical teaching of Adam as our forefather better than Fischer's. I am closest to the position of John Bloom.[9]

MY MODEL IN DETAIL

My model for the correlation between science and Genesis 1 was originally offered in the 1970s as a suggestion for investigation rather than as the last word on the subject. I decided to use literal days rather than ages both to avoid objections from young earth creationists and to reflect the fact that the biblical use of the word "day" is commonly a rather short period of time. I selected the idea that these days would open creative periods (rather than have some other relation to the ages of earth history) for reasons of symmetry, as well as for passages in Scripture that could be understood to indicate that God's rest has not yet started. I admit that my model has a number of problems, but I think these to be less serious than those of the framework model, which I will discuss below.

Some of the objections to my model could be avoided by adopting Hugh Capron's version of the intermittent-day view,[10] in which the six days refer solely to God's commands in creation, rather than to their fulfillment, so that there is no need to correlate order of command with order of fulfillment. I prefer my scheme here as it produces a much higher correlation between Genesis and scientific models of origins.

THE FRAMEWORK HYPOTHESIS

Many evangelicals—both old earth creationists and theistic evolutionists—prefer the framework hypothesis, in which

---

[8]Hugh Ross, "The Mother of Mankind," *Facts & Faith* 2 (1987): 1–2; "Searching for Adam," *Facts & Faith* 10 (1996): 4; Hugh Ross and Sam Conner, "Eve's Secret to Growing Younger," *Facts & Faith* 12 (1998): 1–2.

[9]See John A. Bloom, "On Human Origins: A Survey," *Christian Scholar's Review* 27 (Winter 1997): 181–203.

[10]Dallas E. Cain, "Creation and Capron's Explanatory Interpretation: A Literature Search," *IBRI Research Report* 27 (1986).

the days of Genesis are a literary construct rather than actual days in connection with which the events of creation took place. Days 1 through 3 narrate the formation of the realms of sky, sea-air, and earth; and days 4 through 6 fill each of the realms with creatures in succession.

I have no objection to the claim that some sort of structure like this is actually in the Genesis account. My complaint is with those who use this inferred literary structure as a warrant for rejecting the explicit chronological features of the text. This seems to me to be something like straining out a gnat and swallowing a camel. For here, too, we have an example of "fictitious history." For in the literary-framework-only model, the days of Genesis correspond to nothing that actually happened at creation. At least the young earth and intermittent-day models see the six days as actual days in history, and the day-age model sees the days as actual ages in history. But in the literary-framework view, the days are invented to make a parallel between creation and the human workweek, when no such connection actually existed. As I said earlier, I would like to avoid the idea of fictitious history in harmonizing science and theology, nature and Scripture, if I can.

MIRACLE AND NATURAL LAW

Both the Bible and traditional Christian theology distinguish between miracle and nonmiracle, though there is some uncertainty over what this distinction actually involves. Part of this problem may be seen in the book of Job. Like Job and his friends, we cannot go behind the scenes to know what is going on in the unseen spiritual realm. We don't know what is the direct action of God and what is mediated through various instruments or agents in some way or another.

We do know that God has given moral agents (humans, angels, etc.) some sort of freedom on some level, so that we as persons have responsibility and may be praised or blamed in some real way for our attitudes and actions. Yet even in these actions, God can be spoken of as acting in the same act, even when Satan or the Assyrians are the means he uses to accomplish some result (e.g., on David's census, see 2 Sam. 24:1 and 1 Chron. 21:1; God using the Assyrians, Isa. 10, esp. vv. 5–7, 12–

13, 15–16). The biblical passages that speak of God causing the sun to rise and the rain to fall, therefore, need not be understood as claiming that he does this directly, without the mediation of various physical laws or forces.

In regard to natural law, I am postulating (probably with most Christians who are scientists) some sort of structure or machinery that we call natural law. Like everything else, this structure is created and constantly sustained by God. But I think it has a real creaturely existence, rather than being merely a figure of speech. To what extent it is "autonomous" like a watch or is "driven" like an automobile, either by God directly or by angelic subordinates, we do not know. In either case, God is behind it at every moment in every place, guiding and sustaining its operations.

So what are miracles? To use biblical terminology, they are rare, amazing, powerful, and significant events that seem to override the regular operations of nature and that point to some supernatural power (not always God's) that is here seen to be at work. In the case of creation, I believe God has intervened in this miraculous sense (not denying his continual providential control of all that occurs) to show us he exists and to display something of his power and wisdom.

NATURAL THEOLOGY

I am uneasy with the views of Karl Barth and of Cornelius Van Til, as different as they are, in that they seem to downplay the significance of general revelation and natural theology.

The Bible clearly teaches that general revelation exists. Of course, fallen human beings have a strong tendency to distort the message of general revelation, but I see this as a moral, relational problem (rebellion) rather than the malfunction of our epistemic equipment (disability). In any case, this human problem is operating when we interpret the Bible just as much as when we interpret nature.

Natural theology, then, can be understood in two ways. It can be used to mean what humans typically do learn from general revelation, or what they ought to learn from it. This would be parallel to defining science as what scientists currently believe about some question, versus what they ought to believe, given the data they have, were they handling it fairly. Similarly, in exegetical theology we can make the distinction between what particular theologians actually believe and what they ought to believe, given what the Bible actually teaches.

I maintain the reality of general revelation, from which one can construct a valid natural theology. We as Christians should be encouraging some of our number to be in the forefront of investigating these matters. But whether we do or not, God in his compassion on fallen humanity continues to reveal himself anew as science probes ever deeper with newer technologies.

Chapter Three

THE FULLY GIFTED CREATION

Howard J. Van Till

THE FULLY GIFTED CREATION

*"Theistic Evolution"**

Howard J. Van Till

1. OVERALL POSITION

Personal Position on the Creation-Evolution Controversy

The Beginning of an Answer. To be very candid, I think that the creation-evolution controversy among Christians is the outgrowth of a serious misunderstanding both of the historic Christian doctrine of creation and the scientific concept of evolutionary development. I would even be so bold as to add that the misunderstanding of the historic doctrine of creation may be as widespread within the Christian community as it is outside of it, and that the misunderstanding of the scientific concept of evolution may be as widespread within the scientific community as it is outside of it. If this assessment is correct, then the controversy constitutes a regrettable mistake that must be repaired if the Christian church wishes to be effective in its presentation of the Gospel to a scientifically knowledgeable world in the centuries to come.

Nonetheless, although flawed concepts both of "creation" and "evolution" may be the source of the problem, here we are,

*Although the author of this chapter finds this label to have serious shortcomings, the editors have nonetheless chosen to employ it. The reader is encouraged to take careful note of the reasons, stated within this chapter, why the author asks that his position be known, not as *theistic evolution*, but as *the fully gifted creation perspective.*

engaged in a controversy that continues to cause a division of the Christian community into several camps, each of which is tempted to see itself as superior—either spiritually or intellectually—to all other camps. The fundamental questions at issue concern the character of divine creative activity and the nature of the creation that is the outcome of God's creative action.

What is the best vocabulary to employ in our speech about God's creative work? Is God's creative action best described in a vocabulary that places especial emphasis on episodes of *miraculous intervention* in which God is believed to have imposed new forms on the raw materials that he made in the beginning? Or is it better described in a vocabulary that emphasizes God's *giving of being* to a creation that is richly gifted with the capabilities to organize and transform itself into new forms in the course of time? Is the creation in fact gifted with all of the capabilities necessary to make possible the continuous evolutionary development envisioned by the majority of natural scientists today? Or has the scientific community committed a massive interpretive blunder, and should Christians expect, therefore, that a reexamination of the observational evidence will convincingly discredit the scientific concept of evolution?

Three major camps of Christian positions regarding these questions are presented and evaluated in this book, and I commend the publisher for conceiving of a work in which persons who represent such diverse viewpoints are given an opportunity to present their cases within the same volume. I sincerely hope that the Christian community will become better prepared to discuss the issues of creation and evolution as a consequence of considering the views here presented.

Before I continue my response to the questions provided by the editors of this volume, however, I must devote some time to placing the issues at hand in a larger context. I will also need to spell out the working definitions of several important terms that I will use throughout this chapter. I have often described the creation-evolution controversy as a shouting match that generates "more heat than light"—more hostility than learning. One of the reasons for this unhappy state of affairs is the frequent failure of participants to identify the fundamental questions or to provide clear definitions of key terms.

The Larger Context of the Question. It should be self-evident that the creation-evolution controversy within the Christian community cannot be isolated from the creation-evolution debate between one camp of Christians and another camp of persons claiming to represent the scientific community. To a large number of people, both within and outside of the Christian community, it apparently makes sense to engage in a debate in which a person must choose either creation *or* evolution. Of course, faithful Christians would be expected to choose creation, and anyone who chose evolution would be presumed to stand outside of the Christian community, at least outside of the authentic and faithful portion of it.

The either/or format of the creation-evolution debate is, I believe, one of the most effective factors that has made the discussion of creation and evolution so controversial within the Christian community. If a Christian has been taught that there are only two fundamental perspectives on how the universe got to be as it now is—*creation* and *evolution*—and if he or she is forced to choose between them, how then could a faithful Christian find any credibility in the concept of evolutionary development?

But there are many Christians, especially those of us who are trained in the natural sciences, who feel strongly called to offer a perspective very different from either of the two views ordinarily presented. For me, a Christian who was privileged to be born into a denominational community with a rich theological heritage, this sense of calling arises out of a deep desire to maintain both Christian faithfulness and intellectual integrity. I was taught that maintaining both is not only possible, but also what God desires from me. In this chapter I will present a view in which the Christian doctrine of creation and the scientific concept of evolution are not at all in conflict so that a choice between them becomes unnecessary. In fact, the very idea of an either/or choice between creation and evolution will be seen as wholly inappropriate.

But if you are a person who has been trained to think of creation and evolution as being labels for concepts that stand in radical opposition to each other, the goal of reconciling the two probably looks profoundly impossible. By the end of this chapter, however, I hope that you will see that reaching this goal is not only possible, but immensely worthwhile. Achieving this goal, however, will require some very careful thought. Complex issues demand thoughtful analysis. One aspect of careful thought that will be essential to us is the establishing of a high respect for the precise meanings of important words that will be employed. So, be prepared for a number of carefully stated definitions.

It is no secret that my presentation of a perspective in which the concepts of divine creation and biotic evolution are not treated as adversaries often puts me in a rather unpopular position, especially among conservative Christians. Several years ago I wrote a book titled *The Fourth Day.*[1] In that work I explored the relationship between two portraits of the world's formational history—one based on biblically informed Christian beliefs, and the other based on empirically informed scientific theories. The scientific portrait was represented by descriptions of what we have learned about *cosmic* evolution—the formational history of those inanimate objects and structures of interest to astronomy and cosmology: galaxies and stars, expanding space, and elementary particles. On only a few pages did I make passing reference to the possibility of *biotic* evolution (the formational history of life-forms), noting that I saw no reason to rule it out on either scientific or theological grounds. But guess which pages are most often cited by anxious critics?

Persons who have a strong desire, for whatever reason, to see the discussion cast in the shape of an either/or debate (usually, persons who see their side as the clear winner) do not take kindly to having the debate format discredited. Some preachers of atheism, for example, presume that their no-God message is strengthened by appeal to the scientific evidence favoring biotic evolution and the common ancestry of all life-forms present in the world today. Similarly, some proponents of modern special creationism presume that the best way to demonstrate the truth of Scripture and the need for a Creator-God is to prove that evolution is impossible or that, even if it were remotely possible, it did not actually occur. Although their faith commitments are as different as one could imagine, the two parties of the debate agree in their claim that a simple either/or choice must be made.

Commentary on the Creation-Evolution Debate. Who are the two parties of the debate? When creation and evolution are presented as opponents in a debate, the two positions represented are most often *special creationist theism* and *evolutionary naturalism.* The terms "theism" and "naturalism" both function here as

[1]Howard J. Van Till, *The Fourth Day* (Grand Rapids: Eerdmans, 1986).

labels for worldviews, where by "worldview" I mean a comprehensive set of beliefs about the nature and significance of all reality—the physical universe, the spiritual realm, the world of creatures, the realm of God or of gods, and everything else thought to exist.

By "theism" I mean a worldview founded on a belief in the existence of God. By "naturalism" I mean a worldview founded on the belief that the natural world is all there is to reality, that is, there is no need to consider the existence of God or of gods and the physical universe is presumed to constitute all of reality.[2] Some readers may recall the opening line of Carl Sagan's 1981 *Cosmos* TV series, still available both in book and video format. From the whole series it was clear that by "cosmos" Sagan meant the physical universe, the universe as known by the natural sciences. The naturalistic worldview of the series was clear from the first words of the script: "The Cosmos is all that is or ever was or ever will be."[3]

But both theism and naturalism come in a diversity of specific forms. The theism most often presented in the creation-evolution debate is special creationist Christian theism. It is rightly called both "Christian" and "creationist" because it holds to the historic Christian doctrine of creation—the belief that the one God who is revealed in the Scriptures is the Creator who has given being to the whole universe and who continues to sustain that creation in being. However, among Christians who hold to this fundamental and historic doctrine of creation, there has always been an interesting diversity of pictures of the way in which God's creative activity became manifest in the formational history of the creation.

---

[2]In this chapter I have chosen to use the terms "naturalism" and "naturalistic" only in reference to a comprehensive and atheistic worldview. In other discussions of these issues, however, these terms might be given a much narrower meaning, referring only to the fact that natural science, by the limitations of its competence, formulates theories that refer exclusively to "natural" phenomena (the actions of atoms, molecules, cells, and the like), even though the possibility of "supernatural" action is not at all excluded.

[3]Carl Sagan, *Cosmos* (New York: Random House, 1980), 4. My commentary on the character of this work has been published as "Sagan's Cosmos: Science Education or Religious Theater?" in Howard J. Van Till, Davis A. Young, and Clarence Menninga, *Science Held Hostage: What's Wrong with Creation Science* and *Evolutionism* (Downers Grove, Ill.: InterVarsity Press, 1988), 155–68.

This distinction between "doctrine" and "picture" is, I believe, important to establish and employ in the remainder of our discussion. The historic Christian doctrine of creation is theological in focus (thus not concerned with the details of what might have happened, or when) and calls our attention to the radical difference between the Creator and the creation. The Creator is eternal and self-existent and needs no source of being, but the creation, in contrast, is neither eternal nor self-existent and has being only because God called it into being from nothing "in the beginning" and continues to sustain it in being from moment to moment.

All Christians hold to this historic *doctrine* of creation, but Christians in differing historical and cultural settings have held a rich diversity of *pictures* of the creation's formational history—differing concepts of the way in which God's creative activity (not limited by time, which is a feature of the creation) has brought about (in the course of creaturely time) the remarkably diverse array of physical structures (like atoms, molecules, planets, and the like) and life-forms. All Christians agree that these physical structures and life-forms are creatures—that is, members of the creation—but Christians differ considerably in their judgment of what constitutes the most accurate picture of the formational history of these creatures. In part, this is what the creation-evolution controversy is about. It is largely an argument, among people who hold to the same doctrine of creation, about who has the best picture of the creation's formational history—the best chronicle of how and when creatures came to have their characteristic forms or structures.

The special creationist picture incorporates the claim that the formation of at least some creatures, especially the major "kinds" of living creatures, required acts of "special creation." By an act of "special creation" I mean a creative act of God, performed in the course of time, in which God, acting something like an artisan or craftsman, imposed a new form on the raw materials to which he first gave being—forms that the raw materials could not possibly achieve by using only their own capabilities. This special creationist picture, whether of the young earth (called into being thousands of years ago) or old earth (called into being billions of years ago) variety, is drawn, I believe, primarily from a particular reading of the early chapters of Genesis. I will expand more on this later.

But no matter what the timescale, whether it be thousands or billions of years, the special-creation picture stands in bold contrast to any evolving-creation picture in which God is envisioned as giving being to a creation in an initially unformed state but gifted with all of the capabilities for self-organization and transformation that would be needed to bring about, in time, the full variety of structures and forms that have ever appeared.

The creation-evolution debate, then, is a debate in which the defenders of the creation position have chosen to defend not only the historic doctrine of creation, but also (an extremely significant *also*) one particular picture of creation's formational history—the special creationist picture. This choice necessarily pits them (i.e., the special creationists) against all persons, including many Christians, who find merit in the scientific concept of evolutionary development. However, in the most widely reported form of the creation-evolution debate, the only evolution position offered is the version that is defended by the proponents of *evolutionary naturalism.*

You may recall that earlier I defined naturalism as an atheistic worldview based on the presumption that there is no God, that nature is all there is. By "evolutionary naturalism" I mean a naturalistic worldview presented and defended in such a way that the scientific concept of biotic evolution is made to appear as if it were capable of providing convincing support for the rejection of a Creator-God. This view can be found in a number of well-known books.[4]

Naturalism, in numerous versions, has been around much longer than has the idea of evolutionary development. So what makes these authors think that the scientific concept of biotic evolution can be employed in support of their naturalistic worldview? That's a good question, but first I must clarify what I mean by the *scientific concept of biotic evolution.* By this term I mean the idea—formulated, evaluated, and modified in response to relevant scientific observations and experiments—that all forms of

---

[4]See, for instance, Richard Dawkins, *The Blind Watchmaker* (New York: Norton, 1986); Francis Crick, *The Astonishing Hypothesis: The Scientific Search for the Soul* (New York: Charles Scribner's Sons, 1994); Daniel C. Dennett, *Darwin's Dangerous Idea* (New York: Touchstone, 1995); Jacque Monod, *Chance and Necessity* (New York: Knopf, 1971); and George Gaylord Simpson, *The Meaning of Evolution* (New Haven: Yale University Press, 1949).

life present today have a common biological ancestry and that living systems have all the capabilities necessary to transform (by such processes as adaptation, genetic variation, natural selection, etc.) from the first form of life to the astounding variety of life-forms that have appeared in the course of time.

Note carefully that there is nothing in this scientific concept of biotic evolution that in any way contests the historic Christian doctrine of creation. Our statement of the concept has been crafted in such a way as to limit itself to matters for which the natural sciences have the competence to speak. Nothing in this scientific concept would necessarily incline a person to question the belief that the whole material universe, including the remarkable capabilities of living systems here under consideration, is a creation that has been given being by God.

It is equally important, I believe, to note that nothing in our definition of the scientific concept of evolution would in any way provide support for the fundamental claim of naturalism—that the universe, complete with whatever capabilities it may possess, is self-existent and needs no creator to give it being. Neither does the scientific concept of evolution rule out the idea that the outcome of the universe's formational history was, in a profoundly significant sense, intended for the accomplishment of some divine purpose.

While we're on the issue of purpose, let's look briefly at a common misunderstanding—that *randomness* rules out *purpose*. It is often claimed that the randomness presumed to prevail in the fundamental processes and events of biotic evolution rules out the possibility that evolutionary development could accomplish any preestablished purpose. A simple illustration will be enough to demonstrate the fallacy of this claim. Suppose there were a perfectly honest gambling casino in which no game was rigged—every turn of the cards, every roll of the dice, every cycle of the slot machines, was authentically random. Does that rule out the possibility that the outcome of the casino operation cannot possibly be the expression of some preestablished purpose? Clearly not. In fact, the operators of the casino depend on that very randomness in their computation of the payout rates to insure that they will have gained a handsome profit at the end of the day. Now, if human casino operators can employ random events to accomplish their purposes, could God not do so on a scale far more grand in the formational history of the creation?

Like any concept within the limited domain of scientific competence, the scientific concept of evolution provides no explanation whatsoever for the existence of anything in place of nothing, no ultimate explanation for the particular capabilities that anything possesses, and no way to discern what ultimate purposes the universe's formational history may or may not express. Questions regarding the ultimate source or purpose of the universe's being, which includes its particular capabilities, are extremely important questions, but the natural sciences have nothing to contribute to an answer.

But that brings us back to a question we asked a moment ago: What makes these authors (and other preachers of evolutionary naturalism) think that the scientific concept of biotic evolution can be employed in support of their naturalistic worldview? What makes them think that biological evolution is their ally in the defense of a naturalistic worldview? Ironically, I think a strong case could be made for the thesis that they have been deluded by the either/or format of the creation-evolution debate. They have allowed themselves to think that by defeating the special creationist picture of creation's formational history they have thereby defeated the historic Christian doctrine of creation. Nothing could be farther from the truth. They are living, I believe, in a world of self-delusion. Evolutionary naturalism is in a very fragile state; it can have the appearance of scientific support only in the context of a creation-evolution debate in which the creation position is represented by special creationists of either variety, old earth or young earth. Matters of timetable are of little relevance here.

So, what's the point to be taken from this reflection on the creation-evolution debate? Actually, there are several noteworthy points: (1) it is a debate in which "creation" seldom means the historic *doctrine* of creation, but is most often restricted to mean the special creationist *picture* of God's creative action; (2) it is a debate in which "evolution" is seldom limited to the *scientific concept* of evolution alone, but is most often broadened to represent the *worldview* of evolutionary naturalism; and (3) the debate has therefore functioned to hinder open and fruitful discussion of evolutionary pictures of the creation's formational history within the Christian community.

The challenge before us, then, is how can we Christians, who all hold to the historic doctrine of creation, learn to discuss the merits of an evolutionary picture of the creation's formational history? How can we encourage persons, both within and outside of the Christian community, to set aside the popular misunderstanding that "creation" necessarily implies "special creation"? And how can we encourage persons, both within and outside of the scientific community, to set aside the popular misunderstanding that "evolution" necessarily entails a naturalistic worldview? Having been engaged in this enterprise for many years, I know the difficulty involved; but I think it is imperative that the Christian community put forth the effort and do the careful thinking that is required to converse knowledgeably about these important issues.

Back to the Question. So, then, what is my particular position in regard to the creation-evolution controversy occurring within the Christian community today? If it is a mistake that needs correcting, what perspective would I consider more likely to be correct? If a potentially fruitful discussion has been transformed into a controversy by the ill-conceived creation-evolution debate, how can we recover the benefits of constructive dialogue within our Christian community? And how might we reestablish our ability to reach the scientifically knowledgeable world with the Christian Gospel? The essential elements in my position are these:

1. With Christian people throughout the ages, I hold to the historic and biblically informed Christian doctrine of creation. That is, I believe that the entire universe (everything that is not God) is a creation that has being only because God has given it being, from nothing, and that God continues to sustain it in being from moment to moment. To "create" something is to "give being" to something. If God were to withdraw his creative word, "Let there be," the creation would, I believe, cease to be anything and in its place would be nothing, the same "nothing" that preceded it. In other words, I am a creationist in the full theological sense of the term. I see only two kinds of being: God, who is the Creator (the giver of being), and everything else, which is the creation.
2. What do I see when I look at any of the members of the creation—galaxies, stars, planets, atoms, molecules, cells, living organisms? I see things that have been given a

being that is defined in part by their "creaturely properties"—creatures have properties like size, color, weight, chemical composition, temperature, form, and structure. But the being of creatures is also defined in a very important way by a characteristic set of "creaturely capabilities" to act in particular ways. Atoms, molecules, cells, and organisms, for instance, possess not only properties but also the capabilities to act and interact in a remarkably rich diversity of ways. Those capabilities for acting are essential elements of their being.

3. As a Christian committed to the doctrine of creation, I recognize all of these "creaturely capabilities" as the gifts of being that God has given to his creation. A creature can do no more (nor less) than what God has gifted it with the capabilities to do. And if any one of a creature's capabilities for action were withheld or withdrawn, it would have a different (and less capable) being.
4. From this creationist theological perspective, then, each discovery of a creaturely capability—including every discovery contributed by the natural sciences—provides me with an occasion for giving praise to God for his immeasurable creativity and generosity. In the spirit of this perspective I am inclined to have very high expectations regarding the wealth of capabilities with which God has gifted the creation's being. This high expectation is affirmed each time that the natural sciences come to an awareness of another entry in the list of the creation's capabilities.
5. In part, the creation-evolution controversy is a disagreement concerning the extent of the list of creaturely capabilities with which the creation has been gifted by God. Has the creation been gifted with all of the capabilities that would be necessary to make something like biotic evolution possible? Special creationists are convinced that it has not. I am inclined to believe that it has. I believe that God has so generously gifted the creation with the capabilities for self-organization and transformation that an unbroken line of evolutionary development from nonliving matter to the full array of existing life-forms is not only possible but has in fact taken place.

What would I call such a perspective? Oddly, that presents me with a minor problem. I wish to employ a name that does not carry all of the negative baggage that has come to be associated with some of the more familiar terminology of the creation-evolution debate. And since this book is directed primarily to a Christian audience, I wish also to employ a name that most clearly demonstrates the Christian foundation on which my perspective is built.

Views similar to mine are sometimes identified with the label *theistic evolution*. But that term has some very serious shortcomings. As I see it, it turns the order of importance of divine and creaturely action upside down. Because it appears as the noun, the term *evolution*—which focuses our attention on the natural action of creatures—appears to be the central idea. Meanwhile, by referring to God only in the adjective, *theistic*, the importance of divine creative action seems to be secondary. But that implication would be unacceptable to me.

As a means toward restoring the relative importance of divine and creaturely actions I have sometimes used the label *evolving creation* for my perspective. I think it's a much better term than theistic evolution, but it still has the problem of having to deal with all of the negative attitudes that a majority of Christians have toward anything that even sounds like "evolution." As I have already indicated, the *scientific* concept of evolution, properly defined, does not entail any idea that conflicts with the historic Christian doctrine of creation. The reality is, however, that many persons, both within and outside of the Christian community, and both within and outside of the scientific community, have been led by the rhetoric of the creation-evolution debate to associate the word "evolution" with the worldview of naturalism. That association is, I believe, the result of a serious misunderstanding of both "evolution" and "creation." But even if the association of evolution with naturalism is entirely unfounded, as I believe it is, that association is deeply established in our culture and extremely difficult to correct.[5]

---

[5]Christian critics of evolution are correct in pointing out that many persons, often claiming to speak on behalf of the natural sciences, use the word "evolution" in a way that presumes the truth of naturalism. In effect, they are failing to see, or honestly to acknowledge, the important distinction between the limited "scientific concept of evolution" and the comprehensive worldview of "evolutionary natural-

So, then, what label shall I choose for my concept of a creation that has been equipped by God with all of the capabilities that are necessary to make possible the evolutionary development now envisioned by the natural sciences? For the purposes of the discussion to be carried out in this book, I shall call it *the fully gifted creation perspective*—a vision that recognizes the entire universe as a creation that has, by God's unbounded generosity and unfathomable creativity, been given all of the capabilities for self-organization and transformation necessary to make possible something as humanly incomprehensible as unbroken evolutionary development.

The Integration of Natural Science and Theology

Regarding my general approach to the integration of natural science and Christian theology, I think that in a number of ways science and theology are similar enterprises. Each is engaged in an activity of formulating and evaluating theories. For each enterprise the goal of that theorizing is to develop a system of thought that adequately represents our understanding regarding the character of some portion or aspect of reality. Neither natural science nor Christian theology can claim to deal comprehensively with all the aspects of reality, and neither can claim that their theories capture the fullness of the reality they seek to represent. For instance, as a human enterprise, Christian theorizing regarding the nature of deity will always fail to capture the fullness of God's being. Likewise, scientific theorizing regarding the nature of the physical universe will always fall short of a complete knowledge of the creaturely capabilities that contribute either to the formational history or to the daily operation of the creation. In both cases, a wholesome intellectual humility is appropriate.

Along with these broad similarities there are substantive differences between theology and the natural sciences. One difference regards the principal object of their theorizing. The principal focus of theology's attention is on God. Insofar as we humans (and the rest of the creation) are considered by theology,

---

ism." It is imperative for Christians, therefore, not to mimic that failure, or to accept the false claim that "evolution" and "naturalism" are inseparable, but rather to insist upon the consistent recognition of that distinction.

its concern is with questions regarding the being that God has given to us and where we stand in relationship to God.

The contemporary natural sciences, on the other hand, are focused on the natural world—the creaturely world of atoms, molecules, cells and organisms, the world of planets, stars, galaxies, and the like. The focus of natural science's attention is on learning as much as possible about the properties and capabilities of physical systems, including living organisms, and how systems having those properties and capabilities have behaved in the course of time. Some of this behavior involves the processes of self-organization and transformation that have contributed to the formational history of the universe.

Some writers have chosen to characterize the natural sciences in a way that grants them the right and competence to speak on questions regarding more than the natural or creaturely world, even on questions regarding divine action. I strongly question both the wisdom and accuracy of such a move. I think it greatly exaggerates the competence of natural science. It tempts the preachers of naturalism, for example, into making ridiculously bold assertions, in the name of science, about the ultimate nature of reality. For instance, some of them claim that the universe is self-existent and needs no creator to give it being. Similarly, many writers have presumed that a biblically informed theology has the right and competence to speak on questions regarding specific properties and capabilities of creaturely systems and on questions regarding the particulars of the creation's formational history. I strongly question the wisdom and accuracy of such a presumption.

Both of these moves represent attempts to enlarge the domain of competence for some particular intellectual discipline, either theology or natural science. Such moves are not at all surprising. At this stage in their histories, both theology and the natural sciences are highly professionalized institutions. But institutions, especially ones that have been around for a while, tend to function as self-protective power structures.

For this reason I think we should give questions about the relationship of science and theology a secondary status. Talking about the relationship of two power structures like science and theology is very likely to degenerate into a turf battle—a shouting match in which much energy is wasted in arguing about who should have the right to answer a particular question. As Christians, our primary intellectual concern should be to grow in our knowledge of God, ourselves, and the rest of the creation. In the context of that concern we look to many knowledgeable resources, including both Christian theology and the natural sciences, for answers to our questions. Questions of relationship can then be asked in the specific context of a particular question about the nature of the creation or the ways in which divine action becomes evident. With a particular question in mind, and in the awareness of whatever relevant insights might be contributed by science and theology, we will have to make an informed judgment about the relationship of these two contributions.

Since theology and the natural sciences generally focus their attention on different aspects of reality, authentic conflict is, in my judgment, quite rare. Thus the common presumption that natural science and Christian belief are adversaries engaged in battle must be set aside. Each has the competence to address a list of substantive questions, but the two lists have little, if any, overlap.

That is not to say, however, that what theology and the natural sciences have to offer us can be placed into two noninteracting compartments. That is not the case at all. The issue before us is a case in point. To illustrate that, we need only remind ourselves of some of the central issues in the creation-evolution controversy: How can we best describe the character of divine creative action? By reference to occasional interventions in which a new form is imposed on raw materials that are incapable of attaining that form with their own capabilities? Or by reference to God's giving being to a creation fully equipped with the creaturely capabilities to organize and/or transform itself into a diversity of physical structures and life-forms?

Wording the questions in this way places the emphasis on the theological issue of the character of divine action. But how could a theologian deal responsibly with that issue without giving respectful consideration to what the natural sciences have learned about the creation's formational history? Theology done in isolation from the rest of the intellectual enterprise is unlikely to promote growth in our knowledge of God and even more unlikely to promote growth in our knowledge of his works.

But suppose we had phrased our questions in such a way as to focus attention first on issues falling within the domain of the natural sciences. For instance, we could inquire about the particulars of the formational history of galaxies, or of stars, or of planets, or of the numerous life-forms that have appeared on planet earth in the course of time. Given the discovery that different forms of life have appeared at different times, we could ask about the relationship of those life-forms to one another. Are they all related as parent and offspring? Do all life-forms share a common ancestry? What are the particular creaturely capabilities for self-organization or transformation that may have functioned to accomplish the changes that have occurred? Is that set of capabilities sufficiently robust to make possible the sort of biotic evolution now envisioned by the natural sciences? If so, what are the implications? If we human beings are genealogically related to other life-forms, what happens to our concept of uniqueness as morally responsible beings? Was the appearance of our species intended? Is there any purpose to our existence?

As you can easily see, although our inquiry began with questions entirely within the competence of the natural sciences to address, we ended up with questions that would force a person to go elsewhere for answers. The questions toward the end of our list are clearly questions on which a biblically informed theology has much to contribute—questions regarding human uniqueness, moral responsibility, and purpose. But even if a person were to reject the answers that Christian theology has to offer, that person would still have to go beyond the natural sciences for an answer. Some choose to go to the worldview of naturalism. Unfortunately, many (like the preachers of evolutionary naturalism) do so without realizing that they have left the limited domain of natural science and moved into the much larger domain of a comprehensive worldview. Christians must be alert to the vast difference between natural science and a naturalistic worldview, and they should be prepared to remind proponents of evolutionary naturalism about this difference. Sloppy rhetoric should not go unchallenged.

But what about that question toward the middle of our brief list? Is the set of all creaturely capabilities sufficiently robust to make possible the sort of biotic evolution now envisioned by the natural sciences? This is one of the questions at the heart of the creation-evolution controversy. It is a question that invites a yes or no answer. Those Christians who answer yes will then favor a picture of the creation's formational history with a title like the "fully gifted creation perspective," or "evolving creation" or "theistic evolution." Those Christians who say no will favor a special creationist picture of creation's formational history with a familiar title like "young earth special creationism," or "old earth special creationism," or "progressive creationism," or perhaps the new picture called "intelligent design theory."

On what basis would a Christian come to a yes or no conclusion? An excellent and important question! We will come back to it later. For now the point is this: For the Christian, theology and the natural sciences must be practiced in a way that allows them to be mutually informative enterprises, that is, each must be willing to learn from the other. Each must be respected for the insights or information that it can contribute. Theorizing in either arena must be done in full awareness of what the other can contribute to one's Christian worldview. The particular relationship and interaction of those contributions will have to be evaluated on a case-by-case basis. These relationships are far too complex and diverse to be simplified into just one label or model.

The Role of My View of Integration in the Creation-Evolution Controversy

My vision of natural science and Christian theology functioning as mutually informative enterprises leads me to reject the simplistic either/or format of the creation-evolution debate. In addition to its many other faults, this debate suffers from what is called "the fallacy of many questions." Let me explain briefly what that is.

Think of the whole set of concerns associated with the terms "creation" and "evolution." Imagine making a list of all the questions that are at issue in the debate. Some of those questions fall clearly in the arena of the natural sciences (e.g., physics, astronomy, chemistry, biology, geology, and so forth), while others fall clearly in the arena of theology. In addition, many other categories of questions could be identified—biblical interpretation, philosophy, history, and the like.

In the standard debate format, however, you are presented with only two packages of answers to this long list of categorically diverse questions, and you are told that you must choose one package or the other. But that is a grossly unreasonable request. Differing answers to each question must be examined individually, each on its own merits, not packaged together with particular answers to questions of a vastly different sort.

I have often used a simple illustration to demonstrate the fallacy with which we are here concerned. Suppose I were to hold up a piece of colored paper. I could ask many different questions about this object, including questions of shape and color. Suppose, then, I were to ask you to choose between two packages of answers to these questions. Holding up a square piece of blue paper, I could ask you to tell me if the paper were: (a) blue and circular, or (b) green and square. The fallacy of many questions would soon become evident.

Having found major fault with the simplistic either/or format of the creation-evolution debate and with the controversy that is closely associated with it, I am engaged in developing a perspective that maintains a respect both for Christian theology and natural science that have been performed with competence, professional integrity, and faithfulness to the historic doctrine of creation. In the course of doing so, I must consider a long list of categorically diverse questions, and I must evaluate several possible answers to each. Of these answers, each must be evaluated on its own merits and with due respect for the particular way in which it relates to the other questions at issue.

Sound interesting? It is.

Sound challenging? It is.

Sound easy? It's not. And if the North American Christian community wants a simple and easy solution to the controversy, it will be disappointed. A meaningful solution will come only at the expense of a considerable amount of effort, patience, and perseverance.

2. WHY IT MATTERS

Importance of the Topic

A Divided Church, a Weakened Witness. As Christians we are concerned to present to the world in which we live an effective witness for the Gospel. But if that world hears from us not only our presentation of the Gospel, but also a shouting match among different factions regarding the issues of creation and evolution, what will be their net impression? Having heard us shouting at one another, will they find our invitation to them to join us in worshiping God and living in Christian love as credible?

Must I Reject Scientific Judgment to Accept the Gospel? In the creation-evolution controversy and debate, a major portion of the Christian community is committed to a position that entails the rejection, sometimes expressed in very strident terms, of the scientific concept of evolution—a concept that is judged by the vast majority of natural scientists to be a highly credible theory that gives a remarkably coherent way of accounting for a mountain of observational and experimental data. (Don't forget the important distinction between the *scientific concept of evolution* and the *worldview of evolutionary naturalism*.)

Now, the majority conclusions of the scientific community are not infallible, not even on purely scientific matters. But it is my judgment, and also the judgment of the vast majority of Christian natural scientists that I know personally, that the scientific community is, in fact, soundly warranted in its finding the scientific concept of evolution to be highly credible. Some Christian critics of evolution (often those having no personal experience in the rigors of laboratory research or of scientific theorizing) have expressed the opinion that this judgment should be taken, not as evidence in favor of the scientific concept of evolution, but as evidence for the gullibility of Christian scientists, especially those who teach at Christian colleges. Various poison darts have been thrown by such critics, such as "These folk were brainwashed in their graduate training and are too dull of mind to see it," or "They are intellectually insecure and are willing to buy the acceptance of their secular colleagues in science at any cost," or "They are afraid that they will lose their research funding if they express any criticism of evolution." Pardon my candor, but I cannot label such accusations as anything other than expressions of sophomoric arrogance—ill-founded and vain-spirited accusations that do nothing but amplify the controversy.

One costly outcome of the controversy and debate, then, is a substantially weakened witness for the Gospel to the scientifically literate world in which we live. If scientifically knowledgeable persons are led to believe that in order to accept the

Christian Gospel they must also reject a scientific concept that they have judged, by sound principles of evaluation, to be the best way to account for the relevant observational and experimental evidence, then a monumental stumbling block has, I believe, been placed in their path. I want no part in promoting that false either/or choice between the Christian Gospel and a highly credible scientific concept.

False Dilemmas Can Force Devastating Choices. But the consequences of creating a false dilemma are equally devastating within the church itself. A substantial number of young persons in the Christian community are led to believe that they must make an either/or choice between creation and evolution. Some of them choose to pursue careers in the natural sciences. In the course of their education the majority of them come to recognize the credibility of the scientific concept of evolution. At that point it appears as if they must make a choice in which either option entails a devastating loss: give up their faith or give up intellectual integrity.

The Christian faith, as it was articulated to them by Christian leaders, had come to be associated both with a special creationist picture of creation's formational history and the presumption that the scientific concept of evolution was inseparable from the worldview of naturalism. In fact, they had been told by some of the more strident Christian critics of evolution that the only reason for the general acceptance of the scientific concept of evolution was its usefulness in support of a naturalistic worldview. The empirical evidence, it was claimed, was pathetically weak and could in no way warrant any confidence in the theory of evolution.

For these students, graduate education in the sciences can lead to painful disillusionment. The actual state of affairs turns out to be vastly different from what they had been led to expect. The narrow focus of classroom presentations provided no support for the idea that there was some scientific establishment that desired to brainwash them into accepting a naturalistic worldview. And acquiring a familiarity with the actual data and with the way in which the scientific community formulates and evaluates scientific theories, including the several theories that comprise the scientific concept of evolution, dispelled the claim that the credibility of evolutionary theory was based, not on convincing empirical evidence, but on the corruptive propaganda of naturalism.

What does a Christian student then do? Reject the faith to maintain intellectual integrity? Reject intellectual integrity to maintain the faith? What devastating choices! Both are accompanied by substantial losses. Surely there must be a way to avoid setting young people up for the gut-wrenching pain that accompanies this dilemma.

Broader Cultural and Intellectual Implications

An Ambitious Menu of Intellectual and Cultural Advances. As I have already stated, I view the creation-evolution controversy and debate as being the outgrowth of serious misunderstandings of both the historic Christian doctrine of creation and the scientific concept of evolution. As such, the controversy functions both as a divisive factor within the church and as a factor that seriously impairs its witness to a scientifically knowledgeable world. It must, therefore, be resolved.

Given the misunderstandings from which it proceeds, the resolution of the controversy will, I believe, require a number of substantial changes on the part of a major portion of the Christian community. If accomplished, these changes would, I believe, effect a significant and constructive change in the way Christians engage a number of intellectual issues of the day and the way in which Christian scholarship could be respectfully engaged by the academic community at large. The culture-modifying changes that I see as necessary include the following:

1. The Christian community must resolve to abandon the ill-conceived either/or debate format for discussions on the topics of creation and evolution.
2. It must recover the historic Christian doctrine of creation as a theological commitment that is essential to the Christian faith, but distinct from any particular picture of the creation's formational history.
3. It must be willing to grant the credibility of the scientific concept of evolution as an instructive picture of the creation's formational history and to encourage the scientific community to develop a more comprehensive and detailed understanding of it.

4. It must demonstrate that the Christian doctrine of creation provides a far more substantive foundation for the concept of evolution than does the worldview of naturalism.
5. It must commit itself to encouraging its theologians to aid us in incorporating this concept of creation's formational history into our contemporary articulation of the historic Christian faith.

That's quite an ambitious menu, but let's take a closer look at some of the entries. Since the first two entries have already been addressed, we will begin with the third, which calls for a positive attitude toward the scientific concept of evolution. But before that goal could even be considered to be reachable, two things would have to happen. First, a major portion of the Christian community must become convinced that the Scriptures, particularly the early chapters of Genesis, do not at all require one to accept a special creationist picture of creation's formational history. (I will deal briefly with this biblical matter in a later section of this chapter.) Second, those same persons must come to have a much higher respect for the professional and intellectual integrity of the scientific community than it is now willing to grant.

Consistency as a Goal. The inconsistency of attitudes now displayed by many Christians toward the sciences is very puzzling. Christians ordinarily act in a way that implies a great deal of confidence in the competence and professional integrity of the natural sciences—confidence in their working assumptions, in their methodology, in their gathering and interpreting of data, and in their formulation and evaluation of theories regarding the nature and workings of the universe today. When we become ill, we want to have access to the best of medical science. When we fly on an airplane, we entrust our lives to the aeronautical scientists and engineers that designed it. When we use our computers, we marvel at what the community of computer scientists and engineers have accomplished. When we watch television or talk on a telephone via satellite, we are pleased to take advantage of the remarkable devices that have been developed by communications scientists.

But what happens when that same community of natural scientists, employing the same assumptions and methodology, and applying the same principles of theory formulation and evaluation, devotes some of its attention to reconstructing the universe's formational history? When geologists, chemists, astronomers, and physicists agree that the observational evidence convincingly supports the conclusion that planet earth formed four-and-a-half billion years ago in the context of a universe that came into being approximately fifteen billion years ago, what is the response of the Christian community? Remarkably, a major portion, specifically the young earth special creationist camp, chooses to reject the conclusion reached by the vast majority of natural scientists on the basis of prior beliefs proceeding from a particular reading of the biblical text—a reading that is itself based on one particular set of assumptions and interpretative strategies. For a number of reasons, which I will discuss later, I find this inconsistency of attitude wholly indefensible.

Why should Christians come to accept the credibility of the scientific concept of biotic evolution as an instructive picture of the creation's formational history? If this were some novel or untested concept, or if it represented an unusually speculative view held by a minority of the scientific community, it would be appropriate to present here a carefully articulated defense of its credibility. But the scientific concept of biotic evolution is neither new nor untested, nor is it some sectarian theory held only by a minority of persons working in the life sciences. It is a mature concept considered by nearly all biologists to be firmly established. It is a conclusion entirely consistent with the conclusion reached by physical scientists regarding the formational history of the universe of galaxies, stars, and planets. This is not to say that all important questions have been answered or that we know all of the details concerning what took place in the history of life on earth. That will probably never be the case. But what is the case is that *those persons most knowledgeable about the data and its responsible interpretation are in remarkable agreement regarding the credibility of the general concept of biotic evolution.* That state of affairs is, I believe, very significant, and it would be extraordinarily cavalier for someone outside of biology to dismiss it out of hand.

Another Definition. Point 4 above will probably strike most readers as highly unusual. Who would ever consider the possibility that the scientific concept of evolution could find more support in an environment nourished by the Christian doctrine of creation than in the setting of a naturalistic worldview? This may surprise you, but I find this a very defensible proposition, one that has been sorely neglected by most Christians in the context of the creation-evolution controversy.

As I begin my development of this proposition I find it necessary to introduce some new concepts and terms. These concepts and some of the terminology that I choose to employ may be unfamiliar, but they will not be at all difficult to master. (The more familiar terminology has long been employed with very little success. Why not try a new approach and see if we can break out of the present deadlock?) The first thing I need to define is the concept of the *creation's formational economy*. When I say "economy" do not think of concepts like "thrift" or "frugality." Rather, think of the ways in which we speak of the global *economy*, or of *economic* systems. In that context the term "economy" denotes a complex and interrelated system of resources and capabilities that function to make possible the production and distribution of a vast array of goods and services. Focus especially on the reference to the resources and capabilities that must be operative to make the outcome fruitful.

By the "creation's formational economy" I mean a particular set of resources and capabilities with which the creation has been gifted by God, those resources and capabilities that constitute its being. More specifically, imagine making a list of all of the creation's resources and capabilities that contribute to its ability to organize or transform itself into a diversity of physical structures and life-forms. Among the entries on your list would be the following:

- Fundamental particles called "quarks" have the capability to interact in a way that leads to the formation of things with more familiar names like "proton" and "neutron."
- Protons and neutrons possess the capacities to interact and form the minuscule physical structures we call "atomic nuclei."
- Atomic nuclei have the capabilities to act, interact, and transform in various ways that lead to the formation of different atomic nuclei.
- Atomic nuclei and electrons interact and self-organize in such a way as to produce still higher-level structures called "atoms."

- Atoms have the capability to interact chemically with one another and form a diverse array of molecules; similarly, molecules have the ability to act and interact in ways that lead to the formation of different, often more complex, molecules.
- In the larger context of space, massive clouds of atoms and molecules act and interact to form the immense physical structures of interest to astronomy—galaxies, stars, planets, and the like.
- In the microscopic context once again, biologically important molecules have the capabilities for self-organization into complex molecular assemblies. (In the judgment of the vast majority of scientists, some of these molecular systems achieved the attributes of living systems in the course of the earth's formational history.)
- Living cells have a rich diversity of capabilities for differentiating, developing and functioning as parts of complex multicellular organisms.
- Living organisms possess a comparably rich diversity of capabilities for development, adaptation, and transformation into related life-forms. (In the judgment of nearly all biologists, including the Christian biologists that I know personally, these capabilities are sufficient to make possible the common ancestry of all living creatures.)
- At any one time, a diversity of organisms have the capability to function as members of complex ecosystems within the context of some particular physical environment.

A New Way of Stating the Issue. With this concept of creation's formational economy in place, we are now in a position of being able to restate one of the questions on which the Christian community is deeply divided. It is a question regarding the character of the creation's formational economy. You will probably find my wording of it unusual, but I choose to phrase it in a way that, in my judgment, not only gets to the heart of the issue most quickly, but also helps us to see the way in which the creation-evolution controversy and debate are interrelated.

Here, then, is my version of the question at issue in the controversy/debate: *Is the creation's formational economy sufficiently robust (that is to say, is it equipped with all the necessary capabilities) to make it possible for the creation to organize and transform itself from elementary forms of matter into the full array of physical structures and life-forms that have existed in the course of time?*

This is a question that has only two possible answers. What if we answer no? What would we then be saying about the creation or about the character of God's creative activity? In effect, we would be saying at least the following about the creation: that it was never sufficiently equipped with the capabilities necessary to make possible the sequence of self-organizational and transformational processes now envisioned by the natural sciences. That would eliminate the consideration not only of the modern scientific concept of evolutionary development, but also of Augustine's concept—offered fifteen centuries ago in his commentary *The Literal Meaning of Genesis*—of a creation gifted with the capabilities to form the various kinds of creatures, not in sequence, but side by side. Augustine envisioned a creation initially brought into being in some formless state but gifted, as part of its God-given being, both with the potential for exhibiting a diversity of creaturely forms and with the capabilities for actualizing those forms without any new divine creative acts in the course of time. All creating—that is, all giving of being—was done at the beginning. What took place in time was made possible by the capabilities (Augustine called them "seed principles" or "causal reasons") with which the creation was gifted from the outset.[6]

Another way to state the implications of the no answer is to say that it implies that the formational economy of the creation is marked by the presence of gaps—vacancies formed by the absence of certain capabilities—especially gaps of the sort that would make evolutionary continuity (and also Augustine's side-by-side actualization) impossible. But why would there be gaps in the creation's formational economy? Why would certain creaturely capabilities be absent? In the creationist perspective there is only one broad answer. All creaturely capabilities are

---

[6]Some proponents of the special creationist picture of creation's formational history have presumed that special creationism could rightly be considered one of the "deliverances of the faith," that is, a belief that was universally held in the early Christian community. For the results of an investigation that challenges this common presumption, see Howard J. Van Till, "Basil, Augustine, and the Doctrine of Creation's Functional Integrity," *Science and Christian Belief* 8 (April 1966): 21–38.

God-given. Therefore, if some capability is absent, then it must be the case that God intentionally chose to withhold it.

God is, of course, free to do anything that is consistent with his divine being, and we creatures will never be able fully to comprehend God's choices. Nonetheless, I find it theologically awkward to imagine God choosing at the beginning to withhold certain gifts from the creation, thereby introducing gaps into the creation's formational history—gaps that would later, in the course of time, have to be bridged by acts of special creation.

Of Gap-bridging and Miracles. Recall our working definition of *special creation*—a creative act of God, performed in the course of time, in which God, acting something like an artisan or craftsman, imposes a new form on the raw materials to which he first gave being, forms that the raw materials could not possibly achieve by using only their own capabilities. A Christian who is inclined to favor the special creationist picture of divine creative action must, I believe, think carefully about what is being said about God's creative action. Special creationism, in both its old earth and young earth varieties, entails, I believe, at least the following: (1) that at the beginning God intentionally chose to withhold some particular creaturely capabilities from the creation, thereby introducing gaps into its formational economy; (2) that at a succession of later times (some say over thousands of years, others say billions), these gaps in the creation's formational economy had to be bridged by acts of special creation; and (3) in an act of special creation, God overpowers what he had earlier given being and causes it to do something different from, or beyond, what its creaturely being empowers it to accomplish. Special creationism places considerable emphasis on God's creative action as a display of power and desire to coerce material into assuming forms that it was insufficiently equipped to actualize with its God-given capabilities.

There is a common belief that God's *creative* action necessarily falls in the category of *miraculous* action. I believe, however, that one must distinguish clearly between these two kinds of divine action. As I understand it, a miracle is "an extraordinary act of God performed in the presence of human observers for some specific revelatory or redemptive purpose." Think about all of God's acts that the Scriptures specifically refer to as miracles. Do they not fit this description of miracle? Contrast this concept of miracle with the concept of creation as the giving of being. Note how different these two concepts are. I wish to maintain a belief in both, but in doing so I find it necessary to respect their differences.

Back to the question about whether the creation's formational economy is sufficiently robust to "make it possible for the creation to organize and transform itself from elementary forms of matter into the full array of physical structures and life-forms that have existed in the course of time." What if the answer is yes? What might that imply regarding the character of the creation? A number of answers might be offered here: (1) that, although there are gaps in our limited human knowledge of the creation's formational economy, there are no gaps in the economy itself; (2) that the formational history of the creation is more like a single woven tapestry than a patchwork quilt of pieces sewn together; (3) that Augustine's concept of a creation richly gifted both with potentialities for forms and structures and with the means for achieving them may be the general theme of that tapestry, even if the particular picture of side-by-side actualization failed to be affirmed by scientific investigation; and (4) that the scientific concept of evolution, although it is beyond our limited human ability to comprehend, may provide Christians with the most accurate picture of the creation's formational history yet formulated.

A yes answer to our question also has implications concerning the character of divine creative activity, namely: (1) that the essence of creation is not the imposing of form, like a superhuman craftsman or artisan, but the giving of being, which is a uniquely divine act; (2) that God's creativity, in the thoughtful conceptualization of the creation's being, far exceeds our ability to comprehend it; and (3) that God's generosity in giving the creation such a robust measure of being far exceeds our ability to conceive of it. Hence, in the context of a commitment to the historic Christian doctrine of creation, both the robustness of creation's formational economy and the fruitfulness of its formational history attest to the greatness of the Creator who gives it being.

Think for a moment, then, about how robust the universe's formational economy must be in order that something like cosmic or biotic evolution could be a genuine possibility. How could it be that not only does a "something" exist in place of nothing, but the "something" that exists is gifted with a set of formational capabilities so astoundingly robust that it is able to have actualized the whole array of its structures and life-forms in time? Could anything less than the creativity and generosity of God suffice? Would it actually make sense to posit, as do the preachers of naturalism, that a universe just happens to exist and that it just happens to have an astoundingly robust formational economy? Does naturalism provide anything even closely resembling an explanation? Is it not now abundantly clear that Christian belief in a creator provides a vastly superior framework in which to find the scientific concept of evolution to be a credible possibility?

Think for a moment also about the fruitfulness of the universe's formational history. If our scientific understanding is close to correct, then the whole array of creatures now in existence—both physical structures and life-forms—is the outcome of creatures doing what they had been equipped to do. But consider that outcome. What a remarkable array of creatures! That array includes even us, creatures that are far more than mere survival machines. We are gifted beyond comprehension. We even have the capacity to reflect on who we are and whose we are. We are gifted with the ability to commune with our Creator-Redeemer. We are gifted with the capacities to know right from wrong and to choose the right. How could it be that the outcome of the universe's formational history is so fruitful as to include morally responsible creatures? Could anything less than God's blessing suffice?

What do I mean by "blessing" here? Essentially the same as in a common Christian usage of the term. Suppose you were being wheeled into the hospital operating room for surgery and were still alert enough to offer a brief prayer for God's blessing on this surgery. What would you be asking God to do? To overpower the surgeon's hands or to coerce her into performing surgery differently from the way she was trained and equipped to do? To force the cells of your body into doing things beyond their creaturely capabilities? If that were your prayer, why not skip the surgery and avoid the accompanying pain and expense? No, I think that in asking for God's blessing you would be asking God to act in such a way that the sum of all creaturely actions—the actions of the entire surgical team and of every atom, molecule, cell, and organ of your body—would lead to a fruitful outcome. In this spirit, I see the incredible fruitfulness of the creation's formational history as evidence, loud and clear, of God's uninterrupted blessing. This blessing action is uniquely divine in character; it has no creaturely counterpart. It does not violate the being already given to creatures, but assures that their actions will accomplish a fruitful outcome.

Having dismissed the idea of divine blessing, how could a preacher of naturalism begin to explain the astounding fruitfulness seen in the outcome of the universe's formational history? Would it actually make sense to posit that this represents nothing more significant than a stupendous stroke of luck? Does naturalism provide anything even closely resembling an explanation? Is it not now abundantly clear that Christian belief in a creator provides a vastly superior framework in which to account for the fruitfulness of the universe's formational history?

In his book *The Blind Watchmaker*, biologist Richard Dawkins asserted that, "although atheism might have been *logically* tenable before Darwin, Darwin made it possible to be an intellectually fulfilled atheist."[7] I find that statement entirely unwarranted, nothing more than rhetorical bluster. I would say the exact opposite. If the universe is in fact sufficiently gifted to make evolutionary development possible, and if the outcome of that development includes creatures like us who are able to reflect on what it all signifies, then nothing less than the creativity, generosity, and blessing of God could suffice as the intellectually and spiritually satisfying explanation. Dawkins, along with all other preachers of naturalism, has nothing to offer. In his blusterous rhetoric there is nothing for the Christian to fear.

Christian Responses to the Rhetoric of Naturalism. What is the most common rhetorical challenge hurled at Christians by the preachers of evolutionary naturalism? The essence of the most common challenge is this: *If there are no gaps in the formational economy of the universe, then what need is there for a creator?* The particular way of expressing it will vary from writer to writer, but this is, I believe, the central element in evolutionary naturalism's attack on belief in a creator.

Furthermore, it is ordinarily argued that the natural sciences offer a great deal of evidence that favors what I would call the

---

[7]Richard Dawkins, *The Blind Watchmaker* (New York: Norton, 1986), 6.

"robust formational economy principle"—the expectation that the universe has "the right stuff" (in my vocabulary, "a sufficiently robust formational economy") to make evolution possible.

Christian strategies for responding to this naturalistic challenge fall into two broad types. The most familiar approach in the evangelical Christian community is to accept the challenge as a meaningful one and engage in an effort to demonstrate the presence of substantial gaps in the creation's formational economy. From the alleged presence of these gaps it is inferred that there must have been, therefore, a series of occasional episodes of extraordinary divine action to bridge these gaps. (The labels employed for this extraordinary action include *fiat creation, special creation, miraculous intervention, supernatural intervention, extranatural assembly, injection of information,* and similar terms.) This strategy lies at the heart of both "creation science" (a movement that has for several decades sought to employ its own form of science in support of young earth special creationism) and a more recent movement called "intelligent design theory" (more on this later in the chapter).[8] Both movements seek to demonstrate, by appeal to a reevaluation of observational and experimental evidence, that the scientific concept of evolution is untenable. And, of course, if evolution is discredited, so also is evolutionary naturalism. Case dismissed. Christians win!

The other type of response strategy—the one that I am presenting in this chapter—is to expose the naturalistic challenge as being nothing more than the tired old either/or choice that characterizes the creation-evolution debate, and to argue that the most firm basis for the credibility of the robust formational economy principle is a commitment to the historic Christian doctrine of creation. The naturalistic challenge is based on the familiar premise (one that is deeply embedded in North American culture) that a person has only two choices: either special creationism, with its presumption that there are gaps in the creation's formational economy, or evolutionary naturalism, with its claim that a naturalistic worldview provides an "intellectually satisfying" explanation for the robust formational economy principle.

---

[8]For a description of "creation science" by persons who strongly favor it, see Henry M. Morris and Gary E. Parker, *What Is Creation Science?*, rev. and expanded ed. (El Cajon, Calif.: Master Books, 1987).

It should be obvious that every criticism we have made regarding the either/or format of the creation-evolution debate now also applies with equal force to the naturalistic challenge. That's an important negative point, but the accompanying positive point is, I think, even more important, namely, that the most appropriate and effective response to the naturalistic challenge is to say that if there are no gaps in the formational economy of the universe, then that truly remarkable state of affairs should lead a person to recognize the universe as being a thoughtfully conceptualized and fully gifted creation that has been given its being by an unfathomably creative and generous creator. Without the historic doctrine of creation as its foundation, the robust formational economy principle would represent an incredibly improbable state of affairs. Thus, acknowledging the credibility of that principle provides no evidence against belief in a creator as the giver of being. On the contrary, it provides evidence in favor of that belief.

Why I Reject All Forms of Special Creationism

My reasons for rejecting all forms of special creationism could be the subject of an entire book, but let me offer some brief replies.

Biblical Considerations. I am a firm believer in the biblically informed historic doctrine of creation. However, I am equally firm in my belief that the Scriptures in no way require me to favor or adopt the special creationist picture of the creation's formational history.[9]

Scientific Considerations. Joining forces with old earth special creationists, I reject any form of young earth position because I believe the empirical evidence against it is simply overwhelming. The universe really does provide a wealth of means for arriving at a reliable measure of its age. This is so well established that further discussion of the matter is, I believe, unnecessary. For those persons who nevertheless must do so, more than enough literature is available.[10]

What about the various versions of old earth special creationism, including intelligent design theory? The chief scien-

---

[9]See Howard J. Van Till, *The Fourth Day* (Grand Rapids: Eerdmans, 1986).

[10]See, for instance, Howard J. Van Till, Robert E. Snow, John H. Stek, and Davis A. Young, *Portraits of Creation* (Grand Rapids: Eerdmans, 1990).

tific issue here is whether there is sufficient empirical evidence either for accepting or rejecting the robust formational economy principle. Does the universe have "the right stuff" for making full evolutionary development possible, or are certain key capabilities lacking?

The first thing to remind ourselves of is that it is humanly impossible to *prove,* in the strict logical sense, either a yes or a no answer. To prove either case would first require that we know every process and event that comprised the formational history of the entire universe. Clearly we do not now know that, neither will it ever be possible to know that. Furthermore, to prove a yes answer would require that we are able to demonstrate that these processes and events can be fully accounted for in terms of known creaturely capabilities; and to prove a no answer we would have to demonstrate both that all creaturely capabilities are known and that this set of capabilities is insufficient to account for the universe's formational history. In no way is it possible to know all of this to such a degree as to declare that we have proof for either answer.

For what then must we settle? Since decisive proof is absolutely impossible, we will have to settle for informed judgment. Who should make that judgment? Quite obviously, persons who are sufficiently knowledgeable and equipped to do so. Who are the people most likely to be in that position? Biblical scholars? Theologians? Philosophers? Law professors? People neither trained nor experienced in scientific research?

I presume my point is already clear. Informed scientific judgment is best done by people whose professional training and experience is in the natural sciences. But what should a person outside of the sciences do when there is disagreement among scientists? Good question. If there is deep division within the scientific community with large numbers of comparably qualified scientists on either side of a specific question regarding data interpretation or theory evaluation, then I would recommend suspending judgment. However, if it is a matter of a small number of persons contesting a judgment held by the vast majority of scientists, I would consider the majority position far more likely to be correct. Is the majority judgment ever wrong? Yes, that is possible. However, it should also be pointed out that the correct judgment is almost always generated from within the scientific community, not by a nonscientist from the outside.

Sometimes we Christians are tempted to think that we have access to privileged information, say from the Scriptures. Some Christians go so far as to claim that the Scriptures provide sufficient detail regarding the particulars of the creation's formational history (e.g., its timetable) so that scientific conclusions held with high confidence in the scientific community may rightly be dismissed with little regard for the informed judgment of that community. Personally, I find such claims to be an embarrassment, and they lead many scientists to call into question the intellectual integrity of the Christian community.

Historical and Philosophical Considerations. The history of modern natural science is marked by the accumulation of a greater awareness of the remarkable creaturely capabilities that make up the formational economy of the universe. In its investigation of the universe's formational history, science's practice of assuming the truth of what I call the robust formational economy principle has proven to be one of the most fruitful strategies ever devised for advancing our knowledge of how things got to be the way they are. Furthermore, as we argued earlier, the historic Christian doctrine of creation is far more capable of providing a rational basis for this principle than is naturalism. Why, then, would a Christian want to cling to the idea that it is less than 100 percent true, and that in spite of all of the gifts given to the creation, a small fraction, say 0.001 percent, have been withheld so that evolutionary development is blocked by gaps that must be bridged by occasional acts of special creation?

In all fairness, I have to admit that the absence of gaps cannot be proved. We do not know, nor is it possible to know, all of the processes and events that comprise the creation's formational history. We do not know, nor will we ever know, all of the elements of the creation's formational economy. Neither do we know with certainty whether every process and event in the creation's formational history can be accounted for in terms of the creation's creaturely capabilities. Thus it is *logically* possible that gaps do exist, and therefore it is *logically* possible that these gaps may have been bridged by acts of special creation.

True enough, but suppose I were to ask for a similar judgment, not about the creation's formational history, but about its day-to-day operation. Suppose we were to define creation's operational economy to be the full set of creaturely capabilities that function in the creation's day-to-day operation. Think, for instance, of all the things that atoms, molecules, and cells must be capable of doing in order for us to experience just one day of life in the context of this universe.

Now let's ask a series of questions about this operational economy and about our knowledge of it. Do we have knowledge of every physical, chemical, or biological phenomenon that occurs in the universe during one day of our lives? Do we at this time know all of the creaturely capabilities that contribute to the creation's operational economy? Can we prove whether every phenomenon involving atoms, molecules, and cells during the course of one day can be fully accounted for in terms of the capabilities that comprise the creation's operational economy? Clearly, the answer in each case is no, just as it was for the same series of questions regarding the creation's formational economy and our incomplete knowledge of it.

By now I'm sure you see where I am going. Given this incomplete state of knowledge, are Christians then inclined to posit gaps in the creation's operational economy? Are Christians inclined to presume the necessity of special divine acts in the course of each day to bridge these alleged gaps? Are special creationists also inclined to be "special operationists"? If we Christians are not at all inclined to leap from the recognition of gaps in our knowledge regarding the daily operation of the universe to the special operationist perspective, then why leap from gaps in our knowledge of the creation's formational history to the special creationist perspective? Is this inconsistency warranted?

Theological and Practical Considerations. Special creationism lends itself to engagement in the fruitless creation-evolution debate. In addition to the several shortcomings of the debate already cited, one more deserves mention. It has to do with its upside-down scoring system. Given the way in which the two either/or positions are defined, every scientific discovery of a creaturely capability that makes the robust formational economy principle more likely to be true is credited to the side of evolutionary naturalism. Meanwhile, the truth of Christianity, represented in the debate by special creationism, appears to depend on its ability to prove the presence of gaps in the universe's formational economy. Given the rapid rate at which our scientific knowledge of the universe's formational capabilities is growing, it is no wonder that the preachers of naturalism love this scoring system. What astounds me is that so many Christians have accepted this inverted scale as well.

In essence, what the special creationist perspective grants to the preachers of naturalism, especially in the context of the creation-evolution debate, is ownership of the robust formational economy principle. The either/or format of the debate implies that our Christian belief that the world is a creation would be falsified if the robust formational economy principle is true. Nonsense! The preachers of naturalism should never—*never*—be allowed to presume that they own that principle. That is one of the reasons I argued earlier in this chapter that the historic Christian doctrine of creation provides a far superior basis for the credibility of the robust formational economy principle than naturalism will ever be capable of providing. How utterly tragic, then, that naturalism is so often allowed to claim ownership of the fully gifted universe concept. How ironic that Christians would choose to defend a picture of a less than fully gifted creation.

Because the special creationist position entails the presumption that gaps exist in the creation's formational economy, there is a natural inclination to seek empirical evidence for the presence of those gaps. Once again, note the sad irony of Christians searching for, and hoping to find, evidence of gifts withheld. While the proponents of naturalism are engaged in a positive effort to discover even more of the universe's formational capabilities, some proponents of creation science and of intelligent design are engaged in the negative effort of searching for evidence of formational gifts withheld.

Why this concern to preserve a place (in the gaps) for God to act? Why look so intently for the sort of "special effects" that, as one Christian skeptic of evolution often puts it, "make a difference"? There are at least two problematic features of this strategy. First, it carries the implication, whether intended or not, that God's acts can be performed only where there are gaps in the creation's formational or operational economies. To my knowledge, however, orthodox Christian theology has never limited God's action in that manner. God is free to act in any way that is consistent with his own being. Since God needs no gaps in which to operate, the elimination of gaps in the creation's formational economy places no additional constraints whatsoever on his action. Second, if one wishes to speak of divine action that "makes a difference," then one must posit what would be the case without the presence of divine action. But how would a person know what would have taken place without special divine action? Furthermore, if divine creative action is confined to the bridging of gaps in the creation's formational economy, then the implication is, whether intended or not, that whatever takes place outside of those gaps is not dependent on the continuing moment-by-moment action of God—sustaining and blessing, for instance. These are major problems that must be, but rarely are, addressed by proponents of special creation or intelligent design.

My Personal Journey

I was born into a Christian community that places strong emphasis on being faithful to God in one's beliefs, in one's vocation, and in the wholesome integration of the two. Thus, I was taught, both by word and example, that faith and intellectual pursuits are not adversaries in competition but partners in a cooperative search for understanding. I was brought up in an atmosphere in which high value was placed on intellectual integrity and on the goal of articulating one's Christian theology in a way that enhanced one's understanding in all intellectual arenas, including the natural sciences.

In the home, in the church, and in private Christian schools I received a strong Christian education, for which I am most thankful. Nonetheless, I have always felt that a number of important and interesting questions had never been fully answered. That's good; think of what would happen to intellectual curiosity if that were not the case.

After majoring in physics at Calvin College, I went on to earn a Ph.D. in physics at Michigan State University and, after some postdoctoral research, eventually joined the physics faculty at Calvin, where I have been teaching both physics and astronomy.

With a substantial involvement in both the natural sciences and Christian theology, I observed the resurgent creation-evolution debate with considerable interest. But deep interest soon turned into profound dissatisfaction with the way in which natural science and Christian belief were presented as adversaries in combat. After making increasingly disparaging remarks about the shortcomings of this fruitless debate and its negative impact on the Christian witness to a scientifically knowledgeable world, I had to make a decision: either keep quiet about it or try to do something to improve the situation.

I chose to become a full participant in discussions regarding Christian belief and the natural sciences. Given my training, I proceeded on the assumption that theology and the sciences should function as partners in a cooperative and mutually informative enterprise. For the past several decades I have been actively engaged in writing and speaking on a diversity of issues in the faith-science arena.

Although young earth special creationism was—and still is—present within my denominational community, it never really struck me as something that was either biblically or scientifically defensible. More careful evaluation has considerably strengthened that early impression.

In the 1950s I read the book *Beyond the Atom,* by Calvin College chemistry professor John DeVries, and also Bernard Ramm's more well-known work, *The Christian View of Science and Scripture.* Both of these books argued for the need to take the judgment of the scientific community (which, it should be noted, includes a large number of Christians) into account when formulating one's picture of creation's formational history. The particular picture favored by these authors was some version of old earth special creationism.

Since that time I have become increasingly convinced that the special creation picture of the creation's formational history could not be defended as being either required by Scripture or encouraged by sound scientific judgment. Saying this to the Christian community, especially the conservative evangelical portion of that community, has not been easy or without risk. Nonetheless, I consider it important to continue the effort, and to do so with vigor. Along the way I have encountered a large number of Christians who have provided the encouragement and strength necessary to carry on.

3. PHILOSOPHY OF SCIENCE

How Does My Science Affect My Theology?

In addition to the concept of the natural sciences that is implicit throughout this chapter, let me comment here on two issues regarding scientific methodology: (1) theory evaluation, and (2) methodological naturalism.

How Theories Are Evaluated. At the risk of oversimplification, one of the principal activities of the natural sciences is the formulation and evaluation of theories regarding the way in which some physical system operates or came to be formed. To begin our brief consideration of this enterprise we must first remind ourselves that we will never know everything about any physical system. Therefore, it will always be possible to posit more than one theory to account for what we do know about that system. In the language of philosophy, all scientific theories are "underdetermined by the data."

Contrary to a very common misunderstanding, science never *proves* one theory correct at the exclusion of all other possibilities. Instead, what the scientific community tries to do is to come to an informed consensus on which theory is "best" as judged by a list of theory evaluation criteria. Where does this list come from? From experience. Criteria that are generally fruitful are kept. Others are discarded.

Given a number of theories that might be proposed to account for some phenomenon, these theories are subjected to a number of tough questions like the following: Is the theory in question directly relevant to the system's observable properties and/or behavior? Is it possible for the theory to be falsified by some specific observation? Does the theory make qualitatively and/or quantitatively accurate predictions regarding the observed or inferred behavior of the system? Is the proposed theory internally consistent? Is the proposed theory consistent with other relevant theories that are judged to be highly credible? Is the theory in question limited to just one aspect of a system's behavior, or does it give a coherent account of several phenomena simultaneously? Does the theory unify into one explanatory system what might earlier have been thought to be unrelated phenomena? How well does the theory function to stimulate new programs of fruitful investigation? Does the theory have any attractive aesthetic qualities such as beauty, elegance, simplicity, and the like?[11]

---

[11]For a more detailed look at these theory evaluation criteria, see ch. 2 of Van Till et al., *Science Held Hostage* (n. 3 above), or ch. 5 of Van Till et al., *Portraits of Creation* (n. 10 above).

Is this theory evaluation exercise foolproof? Is it guaranteed to give the "right" answer every time? No, but it works exceedingly well and cannot be dismissed without a very substantial basis for doing so. A person who is skeptical regarding the informed judgment of the scientific community must do far more than simply to assert, "Since mainline science can't provide absolute proof, I am free to reject your theory, even if I am not able to propose a specific substitute that has comparable explanatory power."

Is Science Committed to Naturalism? Let's narrow the field of concern a bit. Suppose that you were a Christian engaged in the formulation and evaluation of theories regarding the formational history of life on planet earth. What kinds of explanatory elements should you be prepared to incorporate into your theory? Should you limit yourself, as the natural sciences do, to "natural" explanations only, that is, to the actions of atoms, molecules, cells, organisms, and the like? Or should a Christian scientist be free to propose the occurrence of extraordinary divine acts (like "special creation" or "intelligent design"') as well, especially in cases for which a specific "natural" explanation cannot be completely specified at this time? Wouldn't this be a good way to show that there is scientific evidence for divine action that has "made a difference"? Shouldn't Christians develop a "theistic science" that is different from "secular science"?

An important and characteristic feature of contemporary scientific methodology is its commitment to what has come to be labeled, especially by Christian critics of evolution, "methodological naturalism." But what exactly is the concept represented by this lengthy term, and is it something that a Christian should welcome or reject?

Unfortunately, the term has been used in such a diversity of ways that it is difficult to pin down its exact meaning. Let me focus on the two principal meanings to illustrate once again the importance of definitions. First, it sometimes stands for the idea that *natural* science, by its own admission, has the competence to propose only theories that appeal to *natural* phenomena. Since natural science has no competence to address questions regarding divine action, whether extraordinary or not, its methodology is limited to dealing exclusively with phenomena that can be accounted for by appeal to the capabilities of atoms, molecules, cells, and the like. Given this understanding, if a Christian wished to propose a theory regarding the formational history of life that did make reference to some form of special creation, that theory would not be a "scientific" theory in the usual sense of the term. Rather, it would be an example of a more comprehensive endeavor that one might wish to call "theistic science." It might be a superior theory; it might not. That would have to be judged on the basis of appropriate criteria, and the natural sciences would have to admit their limited competence in this evaluation.

Defined in this way, methodological naturalism functions mostly to call attention to the limited scope of natural science and to what sort of theorizing it has the competence to perform. Unlike the *worldview* of naturalism, there is nothing in the term "methodological naturalism" that forbids the formulation of more comprehensive theories that appeal to extraordinary divine action. It's just that such theories would need to be identified as being something other than a theory of the *natural* sciences.

If this were the whole story of methodological naturalism, we could be content to treat it as a rather neutral term, neither good nor bad, just a reminder of the limited competence of the natural sciences and the limited scope of the theories that it is able to evaluate. In the context of the creation-evolution controversy, however, the term "methodological naturalism" carries far more rhetorical impact than our first definition would imply. So, what's the second (and, I would add, rhetorically mischievous) meaning? It has a structure something like a layer cake. At the base of this multilayer structure is the presumption that the term "naturalism" (when not qualified by any limiting adjective) would ordinarily refer to the comprehensive naturalistic *worldview*, as has been the consistent practice in this chapter. At the next level, however, there is inserted the implication that the methodology of natural science could only be the outgrowth of that naturalistic worldview. (Why limit scientific theorizing to the consideration of "natural" phenomena only? Because, it is asserted, naturalism says that's the only kind of phenomena there are.) Finally, we come to the top and cosmetically appealing layer: the suggestion that one can take the scientific methodology without its naturalistic foundation and call it "methodological naturalism."

Employed in this rhetorical fashion, the implication is that contemporary scientific methodology is the child of a single parent—the naturalistic worldview. Methodological naturalism (sometimes abbreviated MN) is presented as nothing other than naturalism (N) in disguise. What could possibly lead a Christian to accept this wolf in sheep's clothing? Dull-mindedness? Cowardice? Professional insecurity? Selfish pragmatism? All of these have been offered as explanations.

Or is it possible that some of the loose cannons of the creation-evolution controversy have seriously misunderstood the fundamental theological, philosophical, and scientific issues at stake? Given my training in natural science, and given my heritage of valuing sound theological reflection, I have come to the conclusion, stated at the very beginning of this chapter, that profound misunderstandings are indeed the root cause of the controversy.

One of these misunderstandings concerns what might serve as an adequate basis for contemporary scientific methodology, particularly as it applies to theorizing about the formational history of the universe. At the heart of this methodology is the working expectation that all structures and forms that have ever existed are the outcome of processes and events made possible by a remarkable set of creaturely capabilities for self-organization and transformation. How could this robust formational economy principle be warranted? How could the outcome of the universe's formational history be so astoundingly fruitful? As I argued earlier, the only sufficient warrant I can see for these possibilities is the historic Christian doctrine of creation—the universe has been given its richly gifted being by a creator.

Naturalism is in the awkward predicament of having to assume these possibilities *without anything resembling an adequate basis*. In the heat of the creation-evolution debate, however, Christian critics of evolution have made the serious blunder of failing to capitalize on this state of affairs. Instead of demonstrating the superior ability of the Christian faith to warrant a principle that embodies an exceptionally high view of the creation's capabilities, critics of evolution have chosen to accept the concept of a less-than-fully gifted creation and to engage in a search for gifts withheld.

Am I here arguing, then, for accepting methodological naturalism? No, certainly not as a scientific methodology rooted in a naturalistic worldview. I see naturalism as impotent—unable to generate any convincing basis for presuming the universe to be as capable as it appears to be. I argue here in favor of a scientific methodology that proceeds from a commitment to the historic doctrine of creation. I favor a scientific methodology that sees, in the unfathomable creativity and unbounded generosity of God, the possibility of a creation fully gifted with robust operational and formational economies. Is science committed to naturalism? Not at all. Those whose rhetoric suggests it, whether in their preaching for naturalism or in their preaching against evolution, have failed to dig beneath the surface of methodological similarities into the subsoil of worldview commitments.

Drawing from a number of biblical and theological considerations, I envision a creation brought into being in a relatively formless state, but brimming with awesome potentialities for achieving a rich diversity of forms in the course of time. Drawing also from the vocabulary of the natural sciences, I envision a creation brought into being by God and gifted not only with a rich "potentiality space" of possible structures and forms, but also with the capacities for actualizing these potentialities by means of self-organization into nucleons, atoms, molecules, galaxies, nebulae, stars, planets, and the life-forms that inhabit at least one planet, perhaps more.

For this to be possible, however, the creation's formational economy must be astoundingly robust and gapless, lacking none of the resources or capabilities necessary for accomplishing this organizational and transformational task. This astoundingly robust formational economy is, I believe, a vivid manifestation of the fact that the universe is the product, not of mere accident or happenstance, as naturalism would have it, but of *design*, that is, the universe bears the marks of being the *product of thoughtful conceptualization*. From the Christian perspective, this comes as no surprise because the formational economy of the universe—every creaturely capacity that contributes to it—is a symbol of the Creator's generosity in his gifts of active being.

What About Intelligent Design Theory? Like so many of the key terms in the creation-evolution controversy, the word "design" has many meanings that are seldom distinguished from one another.

References to the idea that the world bears the marks of being "designed" are perhaps most often associated with the name of the eighteenth-century English clergyman, William

Paley. In his classic work, *Natural Theology* (1802), Paley employed the analogy of watch and watchmaker. If a person were to find a watch lying on the ground, he noted, wouldn't that person immediately recognize it as something that had been designed? And would it not then be most reasonable to admit this as evidence for the active existence of a watchmaker who had designed it? So also, perhaps even more so, when we contemplate the remarkable features of some living creature that is fully adapted to the peculiarities of its environment, shouldn't we recognize this creature as being "designed"? And if so, then should we not immediately recognize the need for a "Designer"?

Fully aware of the shortcomings of young earth special creationism, and yet desiring to offer a theistic perspective that would be effective in responding to today's preachers of evolutionary naturalism, some Christians have recently become advocates for an approach they wish to call "intelligent design theory."[12] The basic strategy of the intelligent design movement is this: Select and consider, in the light of information drawn from the natural sciences, specific life-forms and biotic subsystems. Then ask the question, Can one now, with the science of the day (often restricted to some particular conceptual vocabulary, like that of biochemistry, as in Michael Behe's book *Darwin's Black Box*), construct a complete and credible account of how that particular life-form or biotic subsystem first came to be actualized in a Darwinian gradualist fashion? If not, the intelligent design theorists argue, then it must be the outcome of intelligent design, not of mindless, purposeless, naturalistic, evolutionary processes.

What, precisely, does it mean to be "intelligently designed"? Good question! And without a candidly stated definition, confusion is likely to arise. In contemporary usage, to be "designed" means to be *thoughtfully conceptualized*. Design requires the purposeful action of a mind. Think, for instance, of a "design team" that develops the concept of a new automobile model. In this sense of the term, all Christians see the universe as being

---

[12]Representative literature written from this perspective include books by law professor Phillip E. Johnson, *Darwin on Trial* (Downers Grove, Ill.: InterVarsity Press, 1991), and *Reason in the Balance* (Downers Grove, Ill.: InterVarsity Press, 1995); an essay collection edited by philosophy professor J. P. Moreland, *The Creation Hypothesis* (Downers Grove, Ill.: InterVarsity Press, 1994); and the book *Darwin's Black Box* (New York: Free Press, 1996) by biochemist Michael J. Behe. A list of intelligent design proponents would also include Stephen Meyer, Paul Nelson, and William Dembski.

designed. We believe that the being of the creation has been thoughtfully conceptualized by God with clear purposes in mind.

To its proponents, however, the concept of intelligent design entails a second element that makes it essentially the same as Paley's earlier concept, which was based on the *artisan* metaphor. One person, the artisan, did both the conceptualization and the construction of what was intended. Paley's watchmaker, for instance, did both the planning and the fabrication (assembly from crafted parts) of the watch. Paley's Designer (like his watchmaker) was presumed to possess both a mind (to conceptualize, or intend) and the divine equivalent of "hands" (the power to manipulate or coerce raw materials into the intended form).

The modern concept of intelligent design entails not only the idea that some particular life-form (or some aspect of the biological functioning of an organism) has been thoughtfully conceptualized by a mind for a purpose. It also entails the idea that the first assembling of that life-form or biotic subsystem must have been accomplished by some (often unspecified) extranatural agent. All evidence and argumentation favoring the intelligent design conclusion is centered on the claim that this particular species or biological system could not have been assembled for the first time by natural means. Intelligent design means both *thoughtfully conceptualized* and *assembled by an extranatural agent.*

To put the intelligent design thesis in the vocabulary of this chapter, our present failure to comprehend exactly how some particular species or biotic subsystem came first to be formed is taken to be sufficient evidence that its form was imposed by the action of an intelligent designer. Implicit in this thesis is the expectation that there are gaps in the formational economy of the creation that had to be bridged by acts of intelligent design—acts that compensate for the insufficient capabilities of the raw materials. If this is so, then the scientific concept of evolution has been discredited; and if evolution has been discredited, so has evolutionary naturalism. Back to the same either/or line of thought entailed by special creationism. The scientific examples are crafted in a far more sophisticated manner, but the broad features are the same as special creationism.[13]

---

[13]For more on my evaluation of this movement, see my essay review, "Special Creationism in Designer Clothing: A Response to *The Creation Hypothesis*," *Perspectives on Science and Christian Faith* 47 (1995): 123–31.

How This Philosophy of Science Affects My View of Theology and Scripture

First, it should be noted that everyone's reading of the Scriptures and everyone's theology are affected in some way by his or her concept of creation or natural science. My personal goal is to take care that this is done in as self-conscious a manner as possible so that I am in a position to evaluate critically both the process and the outcome.

In the next section of this chapter I will remind the reader that even the scriptural text has been substantively affected by the conceptual vocabularies of its human writers. My wish to respect these historical and cultural influences leads me to resist the temptation to force the biblical text to provide answers to modern questions that can be answered only in a similarly modern conceptual vocabulary. To direct inappropriate questions to the Scriptures is to invite nonsensical answers that would cloud its true and vitally important message.

4. THEOLOGY AND SCRIPTURE

How Does My Theology Affect My Science?

I believe the Scriptures to be divinely inspired and therefore to be "useful for teaching, rebuking, correcting and training in righteousness, so that the man [and woman] of God may be thoroughly equipped for every good work" (2 Tim. 3:16–17). The Scriptures represent, I believe, an authentic account of the divine-human encounter. I believe that God has made and continues to make his presence known to us, and that the Scriptures proceed from a highly important and representative sample of those divine-human encounters. As the verses from 2 Timothy suggest, these writings are especially useful for Christian "training in righteousness."

I must, however, express a considerable degree of discomfort at the way in which numerous Christians have chosen to extrapolate far beyond the words of Scripture itself, heaping onto the words of 2 Timothy many additional humanly constructed claims regarding the biblical text. Some go so far as to assert that the biblical text was intended by God to provide

Christians with inerrant, and therefore binding, information on all sorts of subject matter (including the natural sciences) to which it might appear to be related, even if only tangentially. In its extreme form, this practice of making exaggerated claims regarding the nature and proper use of the biblical text leads to a biblicism bordering on bibliolatry, that is, an inordinate elevation of the status of a historic text, which could lead to the idolization of that text. When a person expends more energy on the strident defense of humanly devised claims about the Book than on living humbly in accord with the actual teachings of the Book, something has gone seriously wrong. Having been the victim of a libelous public attack proceeding in part from that spirit, I believe I have earned the right to comment on the dangers that flow from a profoundly misguided zeal to promote a mischievous concept of Scripture dressed in the garb of a pious attitude toward the Word of God.

Let me offer here just three examples of practices that I find to be especially problematic. Each flows, I believe, from a failure to acknowledge some important feature of the written text itself. The Scriptures must be read and interpreted in a manner that acknowledges their actual character. In my opinion, the violation of this principle and the accompanying placement of unrealistic demands on the biblical text constitutes a serious and prevalent disservice to the community of Christian believers. No matter how good or pious the intentions might be, extensive damage can be done to the Christian witness.

First, the North American Christian community often fails to acknowledge adequately the historical and cultural context of the Scriptures as first written. We are privileged now to have available to us a multitude of modern English translations and versions of the biblical text. There is much to be said in favor of this state of affairs, but there is also an accompanying danger. When we read a text in modern English, it is easy to forget that the original text was written, not only in a vocabulary of *words* provided by a different language, but also in a vocabulary of vastly differing *concepts* that proceed from a very specific historical and cultural setting. Words represent concepts. Concepts are an expression of the culture of some specific time and place. The conceptual vocabulary of modern English is an expression of modern Western culture. It is unavoidable, therefore, that our reading or interpretation of a modern English translation of the Scriptures will be as much influenced by our own conceptual vocabulary as by the conceptual vocabularies of the writers—perhaps even more.

If we wish, therefore, to read the Scriptures for the purpose of "training in righteousness," we are called to pay full respect to the historical and cultural setting in which the text was written. Given the importance of this training, could we justify anything less than that? And what if we wished to go beyond the biblical "training in righteousness"? For instance, what if we wished to bring a number of intellectual questions to the Scriptures? And what if some of these questions were unavoidably framed in the conceptual vocabulary of contemporary Western culture? What if some of these questions were occasioned by the findings of modern natural science?

At the very least we should note two things. First, we are dealing not with the primary "training in righteousness" function of the Scriptures, but with a secondary function—to satisfy our intellectual curiosity. Second, in so doing, we are not only allowed to employ our full intellectual capacities and informational resources, but encouraged to do so. We should not feel compelled to compromise our intellectual integrity in any way. Paying attention to vast differences between our own conceptual vocabulary and the several conceptual vocabularies of the biblical writers would be essential in this endeavor.

How might these principles play out in practice? Let me just point in the direction that the answers to this question will be found, beginning with an example from astronomy. Since the biblical authors had no concept of "galaxy" or of "spatial expansion," it would be absurd to suggest that the word "firmament" in Genesis 1 refers to a "spreading out expanse" that has any connection whatsoever to the modern cosmological theory that galaxies are now moving apart because the amount of space between galaxies in the universe is expanding.[14] Similarly, since the biblical authors had no working concepts of genetic variability, self-organizing molecular systems, genomic phase space, or natural selection, it would strike me as wholly inappropriate to expect the biblical text to offer any unique insights in the eval-

[14]This example is taken from a Christian science textbook for grade six entitled *Observing God's World* (Pensacola: A Beka Book Publication, 1978), 82.

uation of the various specific theories that contribute to the modern scientific concept of biotic evolution. The evaluation of particular theories regarding either spatial expansion or biotic evolution will have to be done in the light of empirical evidence as interpreted by those persons sufficiently knowledgeable and competent to do so. I see no warrant for the idea that the Scriptures could provide the Christian with a shortcut through this complex and intellectually demanding process. To claim otherwise is to risk bringing disrespect toward the Scriptures by making unrealistic claims about them.

A second and closely related problematic practice that we must guard against in regard to our reading and interpretation of Scripture is the frequent failure to acknowledge and appreciate the rich and varied literary artistry found in the Scriptures. Suppose, for instance, that in its original cultural setting some particular portion of Scripture would have been readily recognized as a highly figurative, perhaps even poetic, form of literature. Quite obviously, then, if someone from a different culture were to fail to recognize this literary artistry and proceeded to read it as if it were a sample of matter-of-fact discourse, its reading and interpretation would be badly garbled. No Christian would recommend such a practice, not even the most strict inerrantist.

But suppose we were to take a more relevant (and more difficult) example—Genesis 1. What form of literature is this? It is certainly some sort of narrative (or "story"), but that still covers a lot of literary ground. How is this story form related to the literature of its day? What is the purpose and function of this story in relation to the rest of Scripture? What was it intended to convey to the hearer or reader? Much hangs on the answers to these questions. Within the context of the creation-evolution controversy, two clusters of answers are centered around very different judgments regarding what literary genre (form) this narrative represents.

Clearly, a particular kind of picture of the creation's formational history has been incorporated into the first chapter of the Scriptures. Is this picture itself the message, and is it still normative in the late twentieth century? Is Genesis 1 something like a documentary photograph, or is it more like an artistic portrait? Is Genesis 1 a very concise, matter-of-fact *chronicle* of the formational history of the creation? Should it be read, therefore, not as an example of something written in the form of Ancient Near Eastern artistic literature, but as a relatively artless list of historical particulars of the creation's formational history? Many Christians have been taught to believe this and consequently to draw from the text answers to several questions about what happened when. Young earth special creationists appear to be especially convinced that this is the faithful reading of the text.

Many other Christians are convinced that this is an example of *storied theology.* In this perspective, the *literary form* of Genesis 1 is to be recognized as having great similarity to other available samples of Ancient Near Eastern literature, but the *message conveyed* by this piece of literary artistry is dramatically different. Because of the way in which form and content are closely interrelated, both must be given full consideration in our endeavor to understand the text. In its form, Genesis 1 has the appearance of a portrait of the creation's formational history, an account of how by divine action the cosmos came to be as it is. As such, it is a piece of Ancient Near Eastern *primeval history* literature. But in its concept of the character of deity, and in its concept of who we are and of where we and our world stand in relation to deity, the message of Genesis 1 displays a day-and-night difference from the worldviews of Ancient Near Eastern polytheism.

As the opening lines to a covenant document, Genesis 1 provides us with the answer to one of the most profoundly important questions we can ask: Who is this God who comes to covenant with us and where do we and the powers of this world stand in relation to him? In answer to this question the Scriptures reply, "The God who comes to covenant with you is the One and Only God. There are not many gods, some of them experienced as environmental powers that threaten human existence, but there is only one God and he is the Creator of all else. Everything that is not God is a part of the creation. The powers of this world are not evil deities that seek your demise, but they are members of a creation that has been thoughtfully conceptualized and fully equipped with the capabilities to provide for all creaturely needs." In this view, the one that I would espouse, Genesis 1 provides the elements of the historic doctrine of creation, one of the foundation stones of the Christian faith.[15]

---

[15]For a more extensive development of my reading of the Scriptures, especially of Genesis 1, see the first five chapters of *The Fourth Day* (n. 1 above). For the pro-

Finally, the third problematic practice that we must guard against in our reading and interpretation of Scripture is the frequent failure to acknowledge that the Scriptures, although they do indeed provide the foundation stones for our "training in righteousness," constitute but one of the sources provided for our intellectual growth. If a multiplicity of intellectual resources is provided for the growth of a Christian's knowledge of God and his works, then any practice that so elevates one resource so as to effectively exclude or demean all other sources is, in my opinion, counterproductive. No matter how piously the intent might be stated, the practice of building a picture of creation's formational history from the biblical text alone is simply indefensible. The Reformational slogan *Sola Scriptura* was never meant to be so misconstrued.

How might these considerations apply to special creationist pictures of creation's formational history? The ultimate basis for all such pictures is a particular reading of the biblical text, a reading that begins with the belief—a belief that I find unwarranted—that the literary form of Genesis 1 is not primeval history (a form of "storied theology"), but rather chronicle. If it were not for that reading of the scriptural text, there would not be the widespread belief in special creationism that we see today. All advocacy for a reinterpretation of empirical evidence follows from that prior interpretive commitment. The young earth special creationist picture at least has the merit of attempting to follow a consistent interpretive strategy in the early chapters of Genesis. If it were indeed a chronicle, then one should take the timetable as well as the other pictorial elements as being historical particulars. In order to maintain that timetable, however, the judgment of both the old earth special creationists and practically the entire scientific community must be thrown out. On the other hand, old earth special creationism, by its choice to accept the scientifically derived timetable for cosmic history, is in the exceedingly awkward position of attempting to interpret some of the Genesis narrative's pictorial elements (interpreted as episodes of special creation) as historical particulars but treating the narrative's seven-day timetable as being figurative. I see no convincing basis for this dual interpretive strategy.

---

fessional perspective of an Old Testament scholar, see John Stek's chapter, "What Says the Scripture?" in Van Till et al., *Portraits of Creation* (n. 10 above).

In summary, I believe the Scriptures to be divinely inspired accounts of authentic divine-human encounters, written in the conceptual vocabularies of the human writers, and to have continuing great value for "training in righteousness." In our reading and interpretation of the Scriptures, both for training in righteousness and for intellectual growth, it is imperative that we pay due attention to the text's historical and cultural setting, that we appreciate the richness and variety of the literary artistry found in the Scriptures, and that we acknowledge that the Scriptures constitute but one of many sources provided by God for our growth.

In all of this it is also imperative that we resist the temptation to extrapolate from the biblically warranted concept of "divinely inspired" to the unbiblical concept of "divinely written." The rhetoric associated with the concept of "divinely written" can be offensively arrogant (recall the bumper sticker, "God said it, I believe it, that settles it") and can encourage an unhealthy biblicism bordering on bibliolatry.

The Use of Scripture in the Formulation and Evaluation of Theological Propositions

By this point in the chapter much has already been said that is relevant to the question of how one should use Scripture in the formulation and evaluation of theological propositions. I see theology as an intellectual enterprise, occasioned by the desire to grow in our knowledge of God and his works, that is engaged in the formulation and evaluation of theories, especially theories regarding God and our relationship to him. As such, theology must be done in a way that draws upon all resources of potentially relevant knowledge—the Scriptures, the creation, our continuing experience of the divine-human encounter, and the full array of human life experiences. Although the biblical text may play a special role, it should never be treated as the sole source of theologically relevant knowledge. To build a theology on ancient text alone is to invite the development of a seriously inadequate, insufficiently informed, perhaps even misinformed, theology.

Because theology receives its nourishment from several resources—resources that are themselves active and growing—

theology cannot be static. It will never be completed. Theology must continue to grow as the divine-human experience continues to accumulate and to occur in the context of novel circumstances. Theology must also continue to grow as our knowledge of the character of the creation and of its formational history grows.

I once heard a theologian in my community say that theology was now pretty much done; nothing new should be expected. I don't believe that for a minute. To have a static theology that fails to grow would be a tragedy. In fact, I think Christian theology is now long overdue for a spurt of growth stimulated by our growing knowledge of the creation and its formational history.

One of the tragic consequences of the creation-evolution controversy is that many Christian theologians have been effectively discouraged from the enterprise of rearticulating the faith in light of what the natural sciences have learned about the evolutionary character of creation's historical development. If the sciences are correct in their informed judgment regarding the creation's robust formational economy, what good could possibly come from suppressing theological reflection on this state of affairs? Of what value would a head-in-the-sand theology be? I would encourage the most intellectually gifted of Christian youth to consider the challenge of bringing our theological reflection up-to-date in its engagement of contemporary science.

How This View of Theology and Scripture Affects My View of Science

Given my commitment to the biblically informed, historic Christian doctrine of creation, I am obliged to practice natural science in a manner wholly consistent with that doctrine. One consequence of this is that my presuppositions regarding the character of the universe's operational and formational economies must be consistent with my belief that the universe is a creation that has been given its being by the God attested to in the Scriptures. The scientific methodology that I choose to employ must also be consistent with that belief.

As I have already argued at considerable length, I am inclined by my theological perspective to have high expectations regarding the formational capabilities with which the creation has been gifted. In light of these commitments and expectations, I am not at all surprised to see that the community of natural scientists has found what I call the robust formational economy principle to be an exceedingly fruitful assumption in its endeavor to formulate and evaluate scientific theories regarding the formational history of the universe.

All questions regarding the particulars of the creation's physical properties and capabilities—all of the specific resources and capabilities that comprise its operational and formational economies—are to be addressed to the creation itself and may be investigated by means of the natural sciences. The conceptual vocabularies of the scriptural text are so far removed from this enterprise that a Christian has no basis for presuming that any substantive detail could be derived from the Bible.

5. EPISTEMOLOGY

Resolving Apparent Tensions Between Science and Theology

As a scientist I do hold many scientifically informed and rationally systematized beliefs regarding the physical properties, capabilities, and formational history of the universe. As a Christian I also hold many biblically informed and theologically systematized beliefs about God, about myself, and about the rest of the creation. I believe that the formulation and evaluation of both sets of beliefs must be done thoughtfully, and that each set of beliefs must be allowed to function in the evaluation of the others.

Should either one of these two sets of beliefs consistently be given a higher status, particularly when there appears to be a tension between the two? It is, of course, impossible to give a single answer to this question. Within each set, particular beliefs are held with substantially differing levels of confidence. That is an unavoidable consequence of our being human—we are not omniscient. In the area of scientifically informed beliefs, for instance, I hold the atomic theory of matter with an extremely high degree of confidence, but particular theories regarding the precise sequence of events in a supernova explosion must be held with a greater sense of tentativeness. In the arena of biblically informed theological beliefs, I hold the doctrine of creation without reservation, but I view our reflection on the character of life after resurrection as being more speculative in nature.

Thus, when tension appears, the more confidently held belief, independent of the set from which it comes, is likely to be retained and the other is more likely to be modified. No simple rule can be established ahead of time. Both our knowledge of God and our knowledge of the creation are incomplete and imperfect. Both our interpretation of Scripture and our interpretation of empirical evidence are subject to error. Intellectual humility is appropriate in all cases.

A LETTER TO SUSAN

Dear Susan,

Since your phone call yesterday, I've been thinking a lot about your predicament. You are hearing conflicting claims from persons that you respect. Now you find yourself puzzled about how to make choices among several alternatives. I can't make these choices for you; you will have to do this yourself. What I can do is suggest some basic principles to keep in mind. I include them below:

1. Give yourself sufficient time to evaluate the alternatives carefully. Don't rush and don't allow yourself to be pushed into hasty decisions.

2. Watch out for the fallacy of many questions. Don't accept the simplistic either/or rhetoric of the creation-evolution debate. Deal with each question on its own merits.

3. Use all of your God-given gifts and resources in the spirit of faith seeking understanding. You are gifted with a keen intellect and numerous resources to draw upon as you evaluate what each perspective claims to offer. Don't compromise your intellectual integrity.

4. Listen most intently to people whose training and professional experience prepares them to assist you in making sound judgments on each particular issue. That's what I frequently have to do. On the basis of my professional training in physics I am prepared to make my own evaluation of many theories in cosmology or astrophysics. In that area of scientific

theorizing I find the robust formational economy principle to be sound and fruitful.

In the area of biology, however, I think my best strategy is to draw on the informed judgment of respected Christian colleagues in biology. What I hear from them is that the robust formational economy principle appears to be equally sound and fruitful in their formulation and evaluation of theories about the formational history of life on earth.

5. Remember that the scientific concept of evolution is no enemy to the Christian faith. To recognize the credibility of this concept, as do the vast majority of scientists today, does not in any way tip the scales toward a rejection of the historic Christian doctrine of creation. In fact, if anyone, like your atheist friend Clarice, ever again suggests to you that the possibility of evolution makes naturalism more attractive, here are some questions you should ask her:

a. Clarice, why is there something (e.g., a universe) in place of nothing? Is "nothing" capable of transforming its nothingness into any form of "something"? Do you really expect me to take such a proposition seriously? Get real, Clarice!

Susan, the most fundamental meaning of "create" is to give being. Only a creator can give being to something from nothing. That's one of the reasons for believing that the universe is a creation that was, and still is, given its being by a creator.

b. Okay, Clarice, let's suppose for a moment that, in spite of the fact that it makes no sense to talk of a "nothing" that has the ability to transform itself into a "something," a "something" nonetheless just happens to exist. What's the probability that this "something" not only exists but also just happens to be equipped with a formational economy sufficiently robust to transform itself from some primeval state into a universe with galaxies, stars, planets, atoms, and molecules? Happenstance, you say? Just a great stroke of luck, you say? Do you really expect me to take these answers seriously? Give me a break, Clarice!

Susan, the natural sciences do presume the soundness of what I call the robust formational economy principle. But a scientist's familiarity with that assumption should never lull her or him into overlooking what a remarkable state of affairs that represents. It should send anyone into a state of awe and should force a person to ask, How can this be? How could the being of the universe be so richly gifted? Could anything less than the unfathomable creativity and unlimited generosity of God suffice?

c. Finally, Clarice, even if we could imagine a universe with a rich array of capabilities for self-organization and transformation, who could imagine that the outcome of its formational history would be so astoundingly fruitful as to bring about an unbroken succession of life-forms over time, and that morally responsible creatures like us would eventually appear? Is this nothing more than an amusing accident? Could such a thing come about without thoughtful conceptualization or without purpose? Do you really value yourself so little, and do you really expect me to do the same? Get a life, Clarice! (By the way, if you're actually interested, I've got some friends I'd like you to meet.)

Susan, once again, I think we Christians need to challenge proponents of naturalism much more vigorously and confidently than we ordinarily do. It is easy for them glibly to assume that the universe has what it takes to bring about our formation in the context of biotic evolution. But glib assumptions are not intellectually or spiritually satisfying explanations. How could the outcome of an evolutionary formational history be so fruitful? We are far more than mere survival machines. Could anything less than the blessing of God suffice to explain this?

6. Remember the problems that arise when people fail to honor the actual character of the Scriptures and instead make unrealistic demands on the biblical text. In fact, if anyone, like your biblicist friend Charlie, ever again suggests to you that the Scriptures clearly forbid a belief in the concept of an evolving creation, here are some questions you should ask him:

a. Charlie, in your reading of the Scriptures, have you taken into account the fact that they were written, not only at a time and place very different from here and now, not only in a language of words different from ours, but in a substantially different conceptual vocabulary?

b. Charlie, in your reading of the Scriptures, have you taken into account the fact that they were written in a diversity of literary forms? Have you sufficiently appreciated the richness of the artistry displayed by the biblical text? Do your interpretive principles take this adequately into account?

c. Charlie, when you ask the Scriptures for more than the "training in righteousness" that they do provide, and when you read them for the purpose of stimulating intellectual growth, have you taken into account the fact that God has provided us with multiple resources for our intellectual development? Are you careful to direct to the biblical text only those questions that can be answered in its own conceptual vocabulary? Are you paying adequate attention, for instance, to what has been learned through scientific investigation about the character of the creation and about its formational history?

Susan, don't ever allow anyone to put you on a guilt trip for challenging the special creation picture that seems to dominate the North American Christian community. Other pictures have long been respected among Christians. Give the fully gifted creation perspective a thoughtful evaluation as one that encourages high views of God's creativity and generosity and of the fully gifted character of the creation to which he has given being.

What about that new intelligent design movement? First, it's not really new. In many ways it is similar to the appeal to design made two centuries ago by William Paley, but with some differences. For instance, it is now embellished with more sophisticated scientific examples but, ironically, with substantially less engagement of the relevant theological questions. Second, you should ask the proponents of intelligent design how they expect to overcome the inherent limitations of argumentation that necessarily has the following form: If the scientific community cannot now, within the confines of its limited knowledge and limited conceptual vocabularies, give a complete and convincing account of the formation of life-form X or biotic subsystem Y, then X or Y could have been first assembled only by the coercive action of some unidentified extranatural agent. This is commonly called an "appeal to ignorance," a type of argumentation that rarely convinces.

Susan, you have had a difficult puzzle placed before you. May God bless your best efforts to use all of the resources and capabilities provided for you, and may you come someday to a spiritually and intellectually satisfying solution. And when, by God's grace, you do, perhaps you will be able to tell your friends about your magnified awe for the Creator of heaven and earth, the One who has given being to the fully gifted creation.

RESPONSE TO HOWARD J. VAN TILL

Walter L. Bradley

Dr. Van Till rejects the framing of the whole origins debate as a simplistic choice between creation or evolution, creating as it does (in Van Till's mind) a false dichotomy. Van Till believes that "God has so generously gifted the creation with the capabilities for self-organization and transformation that an unbroken line of evolutionary development from nonliving matter to the full array of existing life-forms is not only possible but has in fact taken place," calling his position the "fully gifted creation perspective." He claims such a position allows him to simultaneously maintain his Christian faith and his intellectual integrity.

Since Van Till's belief that matter is sufficiently "gifted" that it can self-organize and transform itself from simple, innate molecules to highly complex living forms such as *Homo sapiens* cannot be substantiated empirically (or scientifically) at this time, it is difficult to understand why he claims that abandoning the debate helps him maintain his intellectual integrity. It is universally recognized by origin-of-life researchers that the formation of the simplest living cell under realistic, early-earth conditions seems almost impossible to imagine.[1] The Cambrian explosion

[1]C. B. Thaxton, W. L. Bradley, and R. L. Olsen, *The Mystery of Life's Origin: Reassessing Current Theories* (Dallas: Lewis and Stanley, 1992); J. Horgan, "In the Beginning . . . ," *Scientific American*, February 1991, 117; Robert Shapiro, *Origins: A Skeptic's Guide to the Creation of Life in the Universe* (New York: Summit Books, 1986); Leslie E. Orgel, "The Origin of Life on the Earth," *Scientific American*, October 1994,

of all of the major animal phyla in a very short five-million-year time period remains one of the greatest mysteries of origins research today.[2]

In his recent book *Climbing Mount Improbable*,[3] Richard Dawkins, the internationally renowned atheist biologist, correctly acknowledges that unless the back side of "Mount Improbable" has a seamlessly smooth path from bottom to top comprised of many, very small steps, then macroevolution cannot have occurred. Yet, we are far from demonstrating the existence of such a path. Quite to the contrary, biochemist Michael Behe has provided numerous examples at a molecular level of multicomponent systems that cannot function and provide any selective advantage until all component parts evolve independently, a prospect that is incredibly unlikely.[4] Behe calls such systems "irreducibly complex." If Behe's hypothesis proves to be so, then Dawkins's back side of Mount Improbable is in fact found to have many large steps, or discontinuities that cannot be scaled by mutation or natural selection after all. The origin of life to give the first living system may represent the most profound example of irreducible complexity.

It is quite possible to avoid the God-of-the-gaps dilemma that seems to be at the heart of Van Till's concerns by arguing a priori that God is the primary cause who is ultimately responsible for creation. The debate then centers appropriately on how God did it, not whether he did it. Did God bring creation to its current place working exclusively in his patterned way, usually called the laws of nature (which unfortunately implies an autonomy that nature does not have), or using some combination of his patterned ways and occasionally unpatterned ways (what we usually call supernatural or miraculous). Such an approach allows for open inquiry and a search for the truth, rather than abandoning the search for truth in favor of adopting a metaphysical assumption about how God must have done it, despite

77; *The Search for Life's Origins*, prepared by National Research Council Committee on Planetary Biology and Chemical Evolution (Washington, D.C.: National Academy Press, 1990).

[2]Robert F. DeHaan, "Paradoxes in Darwinian Theory Resolved by a Theory of Macro-Development," *Perspectives on Science and Christian Faith* 48 (September 1996): 154.

[3]Richard Dawkins, *Climbing Mount Improbable* (New York: W. W. Norton, 1996).

[4]Michael J. Behe, *Darwin's Black Box* (New York: Free Press, 1996).

significant empirical evidence to the contrary. Van Till's assumption that Christians who are skeptical about chemical and biological evolution are motivated by a certain narrow interpretation of Scripture combined with scientific ignorance is clearly untrue of the so-called intelligent design creationists like me, who, unlike Van Till, are troubled by the lack of sufficient empirical data to support a "fully gifted creation."

Finally, Van Till claims, without any rational justification, that "evolutionary naturalism ... can have the appearance of scientific support only in the context of a creation-evolution debate in which the creation position is represented by special creationists." Evolutionary naturalism is a transparent metaphysical blunder, quite independent of what Christians might believe to be true about creation.

Van Till also argues that creationists insist that non-Christians must reject their belief in evolution as a prerequisite to becoming a Christian. But this is a straw-man argument. No sensible Christian would make such a silly demand. People become Christians by believing the right things about Jesus Christ, quite apart from what they believe about evolution. The same would apply to much more important issues such as the inspiration of Scripture. Likewise, there is no reason why a Christian student cannot be introduced to biblical and scientific information about origins and encouraged not to be dogmatic in his or her thinking about this subject, including uncritically accepting Van Till's theistic evolution, or fully gifted creation.

Van Till's proposed solution is essentially to accept the scientific community's claim of a universe in which natural causes can explain everything, and then attribute this outcome to God's initial imparting of special properties to matter that ultimately, according to Van Till, account for the many complexities of the physical universe and especially living systems. While such an idea may in principle be appealing, in practice it seems to have insurmountable problems that can be easily understood with the following illustration.

Suppose I want to throw a rubber ball from the leaning tower of Pisa in Italy in such a way that it will hit a friend who is walking on the plaza directly below. To be successful, I must predict the exact moment at which the ball will reach the ground and my friend. The fundamental physical law that applies to this physical phenomena is Newton's law of gravitational attraction that in differential calculus takes the following form:

$$F = G\, m_1\, m_2 / r^2 = m_1\, a = m_1\, d^2 h / d\, t^2 \qquad (1)$$

where m_1 and m_2 are the mass of the ball and the mass of the earth, r is the distance from the ball to the center of the earth, and G is the gravity force constant. F is the gravitational force that accelerates the ball. If we note that $a = G\, m_2 / r^2 = -g$, which is 32.2 ft/sec² down, then Equation (1) can be simplified to:

$$d\, h^2 / d\, t^2 = -g \qquad (2)$$

with a solution to this differential equation being

$$h\,(t) = -g\, t^2 / 2 + v_0 t + h_0 \qquad (3)$$

where v_0 and h_0 are the initial velocity of the ball and the height of the tower. The details of the math are unimportant. However, notice what four things are necessary to predict the time at which the ball will reach the ground: (1) the form nature takes, as indicated by Equations 1 and 2 from which one can obtain the solution given in Equation 3; (2) the value of the universal constant G (and the derived constant -g); (3) properties of matter or systems of matter such as the mass of the earth (m_2) and the mass of the ball (m_1) in Equation 1; (4) the initial velocity of the ball when released (v_0), and the height of the tower (h_0).

Note that the universal constant and the properties of matter are distinct from the initial conditions required. In general, physical phenomena in nature, including the origin of physical systems such as life, are determined by (1) the laws of nature; (2) universal constants such as the speed of light (c) or the gravity force constant (G); (3) the fundamental properties of matter such as the mass of the proton and the unit charge and derivative properties of matter such as chemical affinities; and (4) initial conditions that are associated with the arrangement of matter and energy in physical systems, which I will hereafter call information. Each of these components is a necessary part of God's design, and they account for our hospitable home on earth and for the many life-forms found here.

Van Till and I would agree that God is immediately responsible for (1) and (2); however, Van Till would claim that the properties of matter alone (3) can somehow provide the necessary information to account for all of the complex structures found in nature (4). For example, until recently it was widely argued that the properties (3) of amino acid molecules (their respective chemical affinities) are responsible for the very specific sequencing or information (4) that give protein their catalytic properties. However, along with two coauthors, I demonstrated that taken as a whole, the sequencing of amino acid molecules in 250 actual proteins is essentially random.[5] Thus, the chemical properties of amino acids cannot account for the information rich properties of complex protein molecules. There are many examples of similar problems in nature from the incredibly specific initial conditions required for the "Big Bang"[6] to give us a suitable universe for a home to the formation of the complex molecules such as RNA[7] and protein[8] necessary for the first living system. It is this infusion of information that does not seem to be explainable by natural laws and the properties of matter alone. The properties of my ball can never determine its initial position or velocity. In like manner, the properties of matter alone are incapable of providing the necessary information to account for the complexity of the universe or life itself.

The self-organizing tendencies in nature to which Van Till alludes are of two types: (1) those based on the basic properties of matter, with crystals being the best example; and (2) those based on simple initial (or boundary) conditions such as vortices or convective heat flow. There are no good examples of how the properties of matter can produce the very high informational requirements of the Big Bang or the origin of life.

Van Till asserts that the vast majority of scientists believe that biologically important molecules have the capabilities for self-organization into complex forms that have the attributes of living systems, an outcome he would attribute to their God-given properties. However, numerous recent articles in the scientific literature give a very different and much more pessimistic

---

[5]R. A. Kok, J. A. Taylor, and W. L. Bradley, "A Statistical Examination of Self-Ordering of Amino Acids in Protein," *Origins of Life and Evolution of the Biosphere* 18 (1988): 135.

[6]William Lane Craig and Quentin Smith, *Theism, Atheism, and Big Bang Cosmology* (Oxford: Clarendon Press, 1993).

[7]See Orgel in n. 1 above, and R. Shapiro, "Prebiotic Ribose Synthesis: A Critical Analysis," *Origins of Life and Evolution of the Biosphere* 18 (1988): 71.

[8]See n. 1 above.

picture.[9] Why then do many scientists like Van Till believe in a naturalistic origin of life? The most likely explanation is provided by James E. Platt, a professor of biological sciences at the University of Denver. In a recent article in *American Biology Teacher* addressing the way most high school and college textbooks treat the origin of life, he notes:

> These presentations often sound like "just so" stories and are guilty of significant overextrapolation about what we know about the origin of life on this planet. Without dwelling on the details, I will freely admit that the arguments about the origin of life are fraught with difficulties that are generally not treated in introductory texts. I will even go further and—at the risk of incurring the wrath of some of my colleagues—will suggest that the main reasons most scientists believe in a naturalistic explanation of the origin of life is not so much that they find the evidence for particular scenarios all that convincing, but rather that they believe passionately in the principle that the same natural processes which have mediated the evolution of life from simple forms would have operated to produce these simple forms.[10]

It is clear in his repeated references to the beliefs of many scientists about a naturalistic origin of life (and evolution) that Van Till wants to gives his reader the misleading impression that these scientific professionals, as he likes to call them, believe as he does because of the preponderance of the evidence, when in fact their beliefs, like Van Till's, are based on philosophical presuppositions. Similar situations would also attend the Cambrian explosion and the Big Bang cosmology, for example.

Van Till's strong preference for a nature that evolves based on its properties (or giftedness) and his aversion to any "intervention" by God, his gapless economy, borders on a deistic worldview. Deism is the view that God created a universe that operates independently, like a windup clock, in contrast to biblical theism that sees God as Creator and Sustainer, like an electric clock. His description of how he would pray for his surgeon was surprisingly deistic in tone. When I have had surgery, I had

---

[9]Ibid.

[10]James E. Platt, *American Biology Teacher* 55 (May 1993): 264.

no hesitation in praying that God would give my surgeon supernatural skill and insight as needed to give me a maximum outcome. If we see God as responsible for the natural as well as the supernatural (which Van Till acknowledges at one point in his paper), then the view of the supernatural as God intervening in nature is seen to be incorrect, since God is immediately responsible for everything that happens. God is simply doing most things in a patterned way (laws of nature, with universal constants and properties of matter), while occasionally doing some things in an unpatterned way, and not just in so-called redemptive history. Finally, his labeling of a creation where the specification of information for new systems may be necessary as "less than fully gifted" is very pejorative.

In summary, Van Till's goal regarding the maintenance of Christian faith and intellectual integrity is better served by the pursuit of truth through an honest inquiry into the natural world without any of Van Till's presuppositions, allowing God to speak to us through his Word and his world. Only time will tell whether the information gaps are real!

RESPONSE TO HOWARD J. VAN TILL

John Jefferson Davis

Writing from my own perspective of *progressive creationism*,[1] I find much in Dr. Van Till's essay with which I am in agreement. I am in hearty agreement with his view that the principal object of Scripture is not nature considered in itself, with its properties and capabilities, but rather "where we stand in relationship to God" and matters of moral responsibility. I too find that the model of complementarity, rather than conflict or separation, is a more adequate way of understanding the relationship between Christian theology and the natural sciences.

I agree with Van Till that too many creation-evolution discussions have been hampered by dichotomistic, "either/or" assumptions about the means that God may have chosen to create in any given case. He is surely correct when he suggests that each case should be examined individually on its own merits.

His observation that randomness does not necessarily rule out the higher purposes of a rational agent is a valuable one, and is consistent with a biblical doctrine of providence, where God can weave apparently chance events into the divine purpose.[2]

---

[1]This term is used to designate the point of view that the universe and the earth are very old, and that God has created over long periods of time through a variety of means, including both special creative acts and some evolutionary processes. In an earlier generation, this point of view was represented, for example, by Bernard Ramm, *The Christian View of Science and Scripture* (Grand Rapids: Eerdmans, 1954).

[2]For theological reflections on the concept of chance, see D. J. Bartholemew, *God of Chance* (London: SCM Press, 1984).

I would also salute the recognition of human fallibility expressed by Van Till. He is surely correct in recognizing that both our interpretations of nature and our interpretations of Scripture are subject to error, and that intellectual humility "is appropriate in all cases." Heeding this admonition would have prevented much needless acrimony in the history of the Christian church's interaction with the scientific claims.

I would differ, however, with Van Till both in my understanding of the biblical theology of creation and certain empirical aspects of the history of life on earth. I believe the author's statements about the historic doctrine of creation would have benefited from a more detailed analysis of the biblical terminology of creation. A word study of the important Hebrew term *bara*, for example, used forty-nine times in the Old Testament and ten times in Genesis with the sense of "create," implies that through God's command something comes into being that has had no prior existence. The word implies a divine action, a creative work beyond human power. The emphasis is on the *newness* of the created object.[3] My concern theologically is that the author's stress on creation's "functional integrity" and the natural order's "God-given creaturely capacities" gives too much emphasis to the *continuities* of God's creative work in the natural order and too little to the points of *discontinuity* that can manifest the *lordship* and *transcendence* of the Creator over the creation. Another way of expressing this concern is to ask whether Van Till's understanding of creation gives too much emphasis to God's immanence in creation and too little to God's transcendence.

Historically, Christian theology has attempted to recognize this balance between the immanence and transcendence of God in creation through the terms *ordinary providence, extraordinary providence,* and *miracle.* The God of the Bible is free to relate to the natural order in any of these three ways. In ordinary providence, God works immanently through the laws of nature, such as causing grass to grow for the cattle (Ps. 104:14), or creating animals through the ordinary biological processes of gestation (Ps. 104:24, 30). In extraordinary providence God redirects the forces of nature for a redemptive purpose, such as causing a wind to blow

---

[3]Helmer Ringgren, *"bara," Theological Dictionary of the Old Testament*, eds. G. Johannes Botterweck and Helmer Ringgren, trans. John T. Willis (Grand Rapids: Eerdmans, 1975), 2:242–49.

quail from the sea to feed the Israelites during the wilderness wanderings (Num. 11:31). In miracles God transcends or suspends the ordinary laws of nature for a redemptive purpose, illustrated by the floating axhead (2 Kings 6:6), the feeding of the five thousand by Christ, or his bodily resurrection.

My concern is that Van Till's understanding of creation may inadvertently minimize the role of *extraordinary providence* and *miracle* in our understanding of the biblical texts, reducing, in effect, God's creative work to the category of "ordinary providence"—except in the case of the Big Bang origin of the space-time universe. This could amount, theologically, to a *functional deism* in one's view of creation, with God supernaturally involved in creation only at some remote beginning, and only subsequently involved with creation in the role of *sustaining* it. This would seem to be an impoverishment of the richness and variety of the biblically attested ways in which God relates to the natural order, and would seem less robust in providing a check against the encroachments of naturalism. If the natural order is so remarkably endowed with self-organizing properties, then what need is there for the "God hypothesis" at all?

Is there not a danger of so emphasizing creation's functional integrity that the clearly supernatural character of biblical teachings such as the Virgin Birth and the bodily resurrection of Jesus Christ are eroded? It would be helpful to have further clarification from Professor Van Till at these points.

As I examine the record of the history of life on earth, I see more evidence of *discontinuity* than does Van Till. In this limited space I would mention only two prominent examples: (1) the origins of the first living cell ("biogenesis"); and (2) the origins of the major phyla and body forms (the "Cambrian explosion"). In both cases I would see the best explanation of the empirical data in terms of the special creative work of God, rather than merely evolutionary continuity.

Relative to the issue of biogenesis, or the origins of life, while it is true that since the famous experiments of Stanley Miller and Harold Urey in 1953, that amino acids (some of the basic constituents of life) have been readily synthesized in the laboratory, it still remains the case that the "origins of life" question is very far from a laboratory solution. The synthesis of the much more complex nucleic acids such as RNA and DNA, and of the complex proteins necessary for life, under conditions that truly replicate the conditions of the early earth, has not yielded successful results in the more than forty years that have passed since the Miller experiments. The speculative suggestions of current origins-of-life workers such as Stuart Kauffman, who speak of the "self-organizing" properties of matter, remain unverified in the laboratory.[4] I agree with the rather somber and realistic assessment of research in this area by Leslie Orgel, himself a leading researcher in the area, who concluded that the many hypotheses that have been proposed concerning the origin of life are "fragmentary at best," and that the full details of how life emerged "may not be revealed in the near future."[5] I would suggest that Professor Van Till has too easily invoked some concept of a fully gifted creation without adequately wrestling with the detailed problems of the biogenesis question.

Another notable example of discontinuity in the history of life is the "Cambrian explosion" some 570 million years before present, when the major phyla appear in the fossil record without obvious precursors. Complex, multicellular organisms such as the trilobites, corals, and crustaceans appear fully formed in an approximately ten-million-year window of geologic time that has been called the "Big Bang" in the history of life. It is notable that, according to Don A. Eicher and Lee McAlester, "there is no fossil record of the origin of these phyla, for they were already clearly separate and distinct when they first appeared as fossils."[6] The so-called Ediacaran fauna, earlier forms dating from about 640 million years before present, including jellyfishes, soft corals, and segmented worms, did not possess shells or skeletons and do not represent plausible antecedents of the Cambrian phyla.[7]

Trilobites are some of the most extensively studied organisms from the Cambrian record. These crustaceans, with segmented

---

[4]S. A. Kauffman, *The Origins of Order* (New York: Oxford University Press, 1993).

[5]Leslie E. Orgel, "The Origin of Life on the Earth," *Scientific American* (October 1994), 83. Origins-of-life theories and laboratory research are helpfully reviewed and assessed in Charles B. Thaxton, Walter L. Bradley, and Roger L. Olson, *The Mystery of Life's Origin* (New York: Philosophical Library, 1984).

[6]Don L. Eicher and A. Lee McAlester, *History of the Earth* (Englewood Cliffs, N.J.: Prentice-Hall, 1980), 236.

[7]These Ediacaran fauna are described in E. N. K. Clarkson, *Invertebrate Paleontology and Evolution* (London: Allen & Unwin, 1986), 48.

bodies and hard shells somewhat resembling modern horseshoe crabs, are also remarkable for their complex, compound eyes. The eyes of the trilobite, some of which are as complex as those of modern insects, appear abruptly in the fossil record and are the most ancient visual system known in the entire history of life.[8] The fossil record gives no evidence of small Darwinian steps that led to the complex structure of the trilobite eyes.

Van Till's hypothesis of a fully gifted creation does not seem to deal adequately with specific evidences of *discontinuity* in the history of the earth, as in the cases of the origins of life and the Cambrian explosion. I believe that the progressive creation model does better justice both to the biblical and empirical data, recognizing both the elements of continuity and discontinuity in the history of God's creative work. The progressive creation model, when it recognizes the markers of special divine intervention, is not an appeal to a God of the gaps—based merely on human ignorance—but a "best explanation" of *what we actually know* about the history of life and its epochs of novelty and new complexity.

---

[8]H. B. Whittington, *Trilobites* (Woodbridge, U.K.: Boydell Press, 1992), 84–85; and Riccardo Levi-Setti, *Trilobites* (Chicago: University of Chicago Press, 1975), 23ff.

RESPONSE TO HOWARD J. VAN TILL

J. P. Moreland

I am grateful to Professor Van Till for a strong, clear contribution to this discussion. If I were going to adopt some form of theistic evolution, Van Till's would most likely be the one I would embrace because of the careful way it is developed. Still, I remain convinced that theistic evolution is both inadequate in light of all the relevant evidence and is a very dangerous compromise in light of the scientism that characterizes the contemporary intellectual climate and to which Van Till's approach inadvertently contributes. My response focuses on two important issues.

VAN TILL ON THE RELEVANT EXPERTS

Van Till claims that while the majority can be wrong, still, it is most reasonable to base our scientific views on judgments held by the majority of those professionally trained in the natural sciences and when we do, we shall accept the general theory of evolution. When a small number of persons (especially when many of them are from nonscientific fields) contest what the majority of scientists hold, we should go with the majority. Thus, Van Till advises Susan to listen most intently to the majority opinion expressed by professionally trained natural scientists.

Clearly, there is some truth in this advice. But we all know that a group of practitioners in some field can be blinded in one of two ways. First, their training may indoctrinate them so thoroughly into seeing some phenomenon in a certain way, that they simply cannot consider other ways of viewing the data, nor can they hear others who offer alternative interpretations. In short, a homogeneous group can be blinded by the social force of their uniformity and develop closed minds to the suggestion that their views may be inadequate. Second, when the phenomenon in question is at the heart of religious and ethical issues and the majority view does important work for the community of practitioners in justifying their ethical and religious lifestyles, this can blind those practitioners.

Neither of these factors entails that the majority view is false or irrational, but clearly, there are dangers here and the scientific acceptance of evolution fits both criteria. Because this is so, and because it is simply not true that evolutionary theory has been increasingly confirmed by the evidence or fruitfully employed to gather new, empirical discoveries that cannot be explained without evolutionary theory, then there is no inconsistency when people trust their doctors or computer experts to tell them about medicine and software, but do not trust evolutionary scientists to describe the mechanisms that generated living things.

How can we tell if the majority of scientific experts are the relevant authority on the evolutionary question or whether those who dissent (including a minority of scientists and those in fields like law or philosophy outside science) are raising a legitimate alternative? I suggest the following guidelines. For one thing, if Van Till thinks that the majority of current natural scientists have more authority than the dissenters, then he needs to answer two questions: (1) Just exactly what scientific facts are the dissenters getting wrong? (2) Just exactly what special savvy do scientists have that gives them a leg up on interpreting those facts? If Van Till cannot answer these questions, then I suggest his advice to Susan be set aside.

For another thing, the history of science shows that it is often someone outside a major paradigm (e.g., a young scientist and, in rare cases, someone outside science) who has the intellectual distance from the major theory to see its faults and consider alternatives. I think young and old earth advocates are fulfilling that function.

Further, I think that macroevolution is accepted largely for sociological reasons and not rational factors. Specifically, the scientism rampant in culture, coupled with the belief that special creationism is religion rather than science, means that evolution is the only view of origins that can claim the backing of reason. All supposedly extrascientific beliefs must move to the back of the bus and are relegated to the level of private, subjective opinion. If two scientific theories are competing for allegiance, then most professionally trained scientists, at least in principle, would be open to the evidence relevant to the issue at hand. But what happens if one rival theory is a scientific one and the other is not considered a scientific theory at all? If we abandon the scientific theory in favor of the nonscientific one, then given the intellectual hegemony of science, this is tantamount to abandoning reason itself. Given this sociological situation, a logjam of closed-mindedness is created and the acceptance of evolution by the majority of scientists can be explained by the sociology.

Finally, Van Till fails to take his own advice. The vast majority of his article conveys his views about matters in philosophy, theology, and biblical exegesis. Since Van Till is trained in science and not in these other fields, his own advice would, I think, require him to refrain from speaking authoritatively on these topics and, instead, defer to the majority of experts trained in these other fields. Now throughout church history, the vast majority of such experts in the church have held views at odds with Van Till's. What's sauce for the goose is sauce for the gander. If he can speak with some authority about matters exegetical, then why try to silence Phillip Johnson by claiming that a lawyer cannot speak with some authority about matters biological?

My advice to Susan is this: When the above factors are present as they clearly are in the creation-evolution controversy, caveat emptor when it comes to accepting the judgment of the majority of scientists! Susan, if we had followed Van Till's advice in religious studies during the 1960s and 1970s, we would have been required to accept liberal reconstructions of the historical Jesus. In my view, Susan, the majority of religious studies professors searching for the historical Jesus in the last few decades are in the same boat described above as the majority of scientists who accept macroevolution. Thank God that a majority of intelligent dissenters have made available reasonable alternatives in both cases.

VAN TILL ON INTELLIGENT DESIGN

Among Van Till's criticisms of intelligent design theory, two should be mentioned. First, he faults advocates of this approach for reevaluating observational and experimental evidence to show that the scientific concept of evolution is untenable. Further, he claims that this approach is not really new but just a more sophisticated revival of Paley's approach of two centuries ago. Apparently, Van Till means this to imply that current intelligent design theory is just a rehash of an old, failed approach. This is just not true. For one thing, the current intelligent design movement employs a wide variety of design-type arguments (e.g., inference to the best explanation, and Bayesian-type probability arguments), whereas Paley merely employed a standard argument from analogy. Further, the recent intelligent design movement uses very sophisticated arguments (as Van Till acknowledges) and appeals to new types of evidence (e.g., information theory, fine-tuning of the universe). So Van Till's chronological criticism is wide of the mark.

Second, Van Till repeatedly claims that intelligent design advocates base their inference to a designer solely on the basis of an argument from ignorance, that is, from the mere fact that current science cannot explain something, intelligent design advocates inappropriately infer that God is required to explain it. Now, even if this were so, that doesn't mean the argument is a bad one. It all depends on whether there is a lengthy pattern of scientific attempts to deal with the problem in question such that a pattern of explanation has emerged and this pattern is grossly inadequate. If so, the research program in question is in a period of crisis, and critics of the program are within their rights to point this out and advocate a different approach. So it is simplistic to fault all inferences to an explanation based on negative evidence against a rival explanation. Only a case-by-case examination can decide the appropriateness of such an argument form.

However, more importantly, advocates of intelligent design do not base their arguments on negative evidence alone. They claim that we have regular experience of intelligent design and direct assemblage of something by an agent when we examine the artifacts made by human agents. From this we learn certain characteristic marks of the products of intelligent design and direct intentional action (e.g., the combination of a wide variety of different parts that work together for some end; the occurrence of an event that is highly improbable and exhibits nonrandom specificity). If we have theological reasons for expecting the natural world to contain such products, and if we discover products that bear these characteristics, then a design inference may well be justified.

Van Till himself is concerned about persuading naturalists to become Christians and he uses arguments in natural theology to support his efforts (e.g., How can something come from nothing without a divine first cause? How can creation have such highly improbable features of astonishing fruitfulness in creation's formational history and the intricate occurrence of its various capabilities without God giving them as gifts to creation?). Advocates of intelligent design applaud him in this and see no reason to withhold such arguments when it comes to specific episodes (e.g., the origin of life) within the formational history of the creation.

RESPONSE TO HOWARD J. VAN TILL

Vern S. Poythress

Dr. Van Till is right in his most important point, that theistic evolution deserves a respectful hearing. Some solid theologians from previous generations, like B. B. Warfield, have allowed that (possibly with exceptions) evolution might have been the *means* by which God brought into being the present kinds of living things. Van Till is also correct that, if theistic evolution were right, it would be a spectacular display of God's wisdom and power, rather than evidence for naturalism.

Unfortunately, Van Till seriously damages his case by presenting theistic evolution mixed in with defective ideas of his own. Let me point out two main defects.

First, Van Till apparently does not allow exceptions to God's general way of proceeding. He states, "All forms of life present today have a common biological ancestry. . . ," and "An unbroken line of evolutionary development . . . has in fact taken place." Van Till would have been better off to allow exceptions. Disallowing exceptions is in tension with his belief in the possibility of miracles and his view that science should be limited and tentative in its claims. Worse, Van Till's view contradicts Genesis 2. Genesis 2:21–22 says that "he [God] took one of the man's ribs and closed up the place with flesh. Then the LORD God made a woman from the rib he had taken out of the man. . . ." (The Hebrew word for "rib" can also sometimes mean "side," "plank," or "side-chamber." The context here shows that it is a rib or a piece from the man's side, since afterward God "closed up the place with flesh.") The Bible indicates not only that God made Eve, but *how* he went about it. He did not use a normal conception and birth.

God works in whatever ways he chooses. We must allow that he could create Adam and Eve by different means than what he uses with other creatures. This key admission destroys the dogmatism with which mainstream scientists typically set forth sweeping evolutionary claims. Moreover, allowing exceptions brings theistic evolution into proximity with a special creation position. Consider that in the one case where the Bible gives the most detailed information about *how* God created, namely, with the creation of Eve, it sets forth an exceptional means. This fact suggests that perhaps God created most *other* major groups of living things by unusual means. The difference between theistic evolution and special creation then boils down to a quantitative debate about how many exceptions there were to normal reproductive processes.

The second defect is that Van Till's view of God's interaction with the world promotes deism, not a Trinitarian biblical view. Perhaps he did not intend it, but most readers will understand him this way. In a deistic view, God produces the original starting point of creation, but is thereafter fundamentally uninvolved. God "winds up the clock" in an original act of creation; afterward, the clock runs "by itself." By contrast, the Bible indicates that God is *constantly* involved in controlling, directing, speaking, and interacting with his creatures: "He makes grass grow for the cattle" (Ps. 104:14); "You bring darkness, it becomes night" (104:20); "He covers the sky with clouds" (147:8); "He spreads the snow like wool" (147:16). The Bible particularly focuses on God's interaction with human beings: "The LORD sent Nathan to David" (2 Sam. 12:1); "The king's heart is in the hand of the LORD; he directs it like a watercourse wherever he pleases" (Prov. 21:1). One could multiply the passages.

In these actions God is so consistent and so in harmony with himself that we can sometimes describe what he does by so-called scientific laws. But there are limits. The laws are approximate human guesses; they are incomplete; they typically focus on only one aspect (mechanical, geological, biological, etc.) rather than the larger picture; and they may have exceptions when God decides for his own reasons to vary the pattern. The original acts of creation, the life of Christ, and the second coming of Christ are such variations. They involve events of spectacular newness, so dogmatic assumptions of mechanical continuity with the present are singularly inappropriate.

Unfortunately, Van Till does not enter into this biblical teaching. And so his exposition leaves readers with the impression that one must choose between two bad alternatives: (1) Either God "intervenes" into a mechanism in order to "coerce" it contrary to its true structure (God of the gaps); or (2) God stands outside the mechanism and lets it work simply "on its own" (a recipe for deism). Van Till's description falsely suggests that God's action would involve disruption "from outside," and thereby would imply a regrettable deficiency in the goodness and completeness of the created order. He made this mistake because he used a defective picture of God. God is not an artisan or craftsman disrupting a self-contained mechanism *from outside.* Rather, he is omnipresent, present "inside" (Jer. 23:24). He is Father and personal agent, constantly active up to the present (Pss. 139:13; 104:2–3, 30; Jer. 18:1–10; Matt. 6:25–34; John 5:17; etc.).

Van Till obscures the fundamental issues when he speaks about the "properties" and "capabilities" of created things. What does he mean? Most people will understand these to refer to descriptive properties and operations in conformity with various known scientific laws. These laws are "methodologically naturalistic," and so never rise beyond the picture of the universe as mechanism. Things are controlled by impersonal forces and mindless interactions. God may have set up the whole universe and exercised his wisdom. But what we have afterwards is self-contained. God has delivered the capabilities over to creation as a once-for-all gift of the distant past. Now the creation exercises these capabilities and God has only to ensure that everything remains in existence.

By contrast, within a biblical view, the capabilities can only be a shorthand for God's decrees and specifications and actions, governing what happens in the world, day by day. Whatever happens, whether in conformity with modern scientific expectations or not, is expressive of God's complete control over his world.

Look at Van Till's own statement: "A creature can do no more (nor less) than what God has gifted it with the capabilities to do." Within a biblical view, this statement is a truism. A creature can do only what God specifies (Lam. 3:37–38). But we cannot impose limits on what God specifies. Jesus turned water into wine (John 2:1–11); Elisha caused an axhead to float (2 Kings 6:6); the body of Jesus is alive from the dead and transformed (1 Cor. 15). All these events *by definition* are realizations of God-given "capabilities," if that term is only shorthand for God's possible plans and actions.

But of course, this key word "capabilities" soon slips back into a different and opposite meaning. "Capabilities" then means *only* those possibilities explicable within the bounds of methodological naturalism, when it is used to analyze the regularities observable today. The natural regularities from today end up setting artificial bounds for what God does.

One final point. On the question of evolution, I do not endorse Van Till's idea of simply relying on scientific specialists to determine the most viable theory. For several reasons current scientific training seriously biases people toward expecting wholly naturalistic explanations. No intellectual dishonesty need be involved, but only the absorption of accepted assumptions and interpretations, and the lack of an attractive alternative that would fit within the same framework of assumptions. For the reader who wishes to explore more about this, I suggest reading Michael Denton's book *Evolution: A Theory in Crisis* (Bethesda, Md.: Adler & Adler, 1985).

In sum, the deistic tendency in Van Till is not acceptable. Theistic evolution is a possibility theologically, if one allows exceptions. But it is unlikely on scientific grounds.

CONCLUSION

Howard J. Van Till

I suppose I should be mildly grateful for Professor Moreland's statement that if he were to adopt some form of theistic evolution, my version would be his choice. However, having made very clear in my chapter that I find the label "theistic evolution" not only inadequate but misleading, I am not particularly amused by his choice to revert to that rhetorically offensive epithet as a label for my perspective. The thrust of my chapter was not to Christianize evolution, but to celebrate the generosity of the Creator and the giftedness of the creation.

In my advice to Susan I suggested that a person is well advised to respect the informed judgment of the professional scientific community with regard to the evaluation of scientific theories. Professor Moreland's response was that although this would ordinarily be sound advice, evolutionary theorizing should be treated as an exception to that generally good rule. Why? Because, thinks Moreland, "macroevolution is accepted largely for sociological reasons and not rational factors."

Many persons in the Christian community apparently wish Moreland's judgment on this to be a true and fair reading of the way things are. I do not believe it for a moment. Having interacted for more than three decades with a large number of Christian biologists, I am wholly convinced that scientific theory evaluation is done with just as much competence, honesty, and rationality in the arena of biotic evolution as in any other area of natural science. The idea that the judgment of the scientific community—a community that includes a substantial number of Christians, by the way—can here be dismissed as a mere sociological distortion is, I believe, an attractive illusion that must be conclusively dispelled. This illusion serves to reassure those persons who wish to maintain some form of special creationism in spite of scientific judgment, but it is, I believe, an illusion nonetheless.

Moreland is, as I expected he would be, critical of my judgment regarding the intelligent design movement. Rejecting my comparison of it with Paley's approach, Moreland notes that the current movement employs far more sophisticated arguments and appeals to new types of evidence, such as the "fine-tuning of the universe."

Moreland's appeal to the fine-tuning of the universe puzzles me, however. Why? Because it illustrates exactly the opposite of what he intends. Let me offer a very brief explanation. Modern cosmological theorizing about the formational history of space, of the elements, and of galaxies, stars, and planets has led us to the realization that very small changes in the values of any of several properties of the universe would have prevented our arrival on the scene. Minuscule modifications in the rate of cosmic expansion, the strength of the gravitational force, the speed of light, or even the placement of nuclear energy levels, for example, would so drastically have altered the formational history of the universe that no form of life, certainly not our form, would have come to be actualized. In the standard parlance, the universe appears to be exquisitely fine-tuned for our appearance.

I heartily agree with that line of thought. The finely tuned character of the nature of the universe does indeed lead me to conclude that it is the product, not of happenstance, but of thoughtful conceptualization. In other words, the universe is a creation that has been conceived by an incomprehensibly creative Mind. But here's the rub: The fine-tuning to which Moreland calls attention is necessary only in the context of a formational history that is constrained to be consistent with what I have called the robust formational economy principle. The fine-tuning of the properties and capabilities of the universe is evidence, not of the need for episodes of extranatural assembly (a basic feature of intelligent design), but of the idea that the universe is a creation that has been fully gifted with a robust and gapless formational economy. Fine-tuning eliminates the need for occasional, form-imposing interventions.

Moreland judges that the argument for intelligent design can be strengthened beyond the standard appeal to negative evidence (what I called an appeal to ignorance) by the positive comparison of selected creaturely systems to artifacts that we know to have been both designed *and assembled* by human agents, but that analogy has, I believe, a serious flaw. Human artisans are limited in ways that the Creator is not. Human artisans are limited to acts of rearranging materials at hand. We cannot *give being* to new substances in the uniquely divine manner in which the Creator has given being to the creation.

Neither can we humans foresee the full range of potentialities in the actions of the substances to which the Creator has given being. That's why we continue to do scientific investigation—to discover more of those creaturely capabilities and potentialities. And to have expectations as high as those entailed by the robust formational economy principle seems wholly appropriate. For Moreland to call this fully gifted creation perspective a "very dangerous compromise in light of the scientism that characterizes the contemporary intellectual climate" is, I believe, grossly unfair to the position I presented in my chapter.

Like Moreland, Professor Poythress begins his response to my chapter with what superficially looks like a compliment: "Dr. Van Till is right in his most important point, that theistic evolution deserves a respectful hearing." Once again, however, the malodorous label "theistic evolution" is used to name my perspective. Not once has my own designation of it as a "fully gifted creation perspective" been employed. Even more seriously, Poythress has clearly missed the central point of my chapter. My principal point was to call attention, not to the viability of any particular scientific theory, but to the unlimited generosity and unfathomable creativity of the Creator.

In the opinion of Poythress, one of two serious defects of my view is that it "contradicts Genesis 2" regarding the specific manner in which the first woman came to be. I presume that he meant to say that it contradicts his *preferred interpretation* of Genesis 2. That would indeed have been an accurate statement. In agreement with numerous biblical scholars in the evangelical

Christian community, I do read the early chapters of Genesis as being literature that conveys profoundly important theological truths in a literary style that is more like an artistic portrait than a documentary photograph.

Commenting on whether the human author of Genesis 2 intended a literal reading of the story of Eve being crafted from the side of Adam, Henri Blocher says, "But the presence of one or several word-plays casts doubt on any literal intention on the author's part; they reveal an author who is in no way naïve, but who uses naïve language for calculated effects."[1] Similarly, Gordon J. Wenham comments that, "Indeed, the whole account of woman's creation has a poetic flavor: it is certainly mistaken to read it as an account of a clinical operation or as an attempt to explain some feature of a man's anatomy."[2] Speaking more generally about his reading of the early chapters of Genesis, John R. W. Stott notes that "my own position is to accept the historicity of Adam and Eve, but to remain agnostic about some of the details of the story...."[3] Now, Professor Poythress is entitled to hold a differing interpretation, but for him simply to assert that his particular reading is the only acceptable one is to neglect the diversity that characterizes the Christian community and to allow personal preference to substitute for intellectual modesty and candor.

The second objection voiced by Poythress (also by Bradley and Davis) is that my view "promotes deism," a perspective that envisions a distant God who neither acts in nor interacts with the creation. This is a common misunderstanding of my view, and I am curious to know why so many Christians are inclined to associate a fully gifted creation with an inactive God. Why, for instance, would Poythress, Bradley, and Davis all express a fear that my position might invite and give comfort to deism? Is it the case that if there are not gaps in the formational economy of the creation, then God must be inactive in it? Not at all, and I think the quickest way to dispel that mistaken inference is to ask the following question: Has orthodox Christian theology ever

---

[1]Henri Blocher, *In the Beginning* (Downers Grove, Ill.: InterVarsity Press, 1984), 98–99.

[2]Gordon J. Wenham, *Genesis 1–15*, in *Word Biblical Commentary*, 2 vols. (Waco: Word, 1987), 1:69.

[3]John R. W. Stott, *Understanding the Bible* (Grand Rapids: Zondervan, 1979), 233.

suggested that God is able and/or willing to act in the world only within gaps in either the formational economy or the operational economy of the creation?

To the best of my knowledge the answer is a resounding, No! Therefore, if the presence of such gaps is *not* required to "make room" for divine action, then the absence of such gaps is no loss whatsoever. End of story.

From the vantage point of believing that God gave being to a creation full-equipped with a robust formational economy, God is still as free as ever to act in any way that is consistent with his nature and will. The full-gifted creation, complete with a gapless formational economy, does not in any way hinder God from acting as God wills to act. As I have said on numerous occasions, the question at issue is not, Does God act in or interact with the creation? Rather, the question is, What is the character of the creation in which God acts and with which God interacts?

Professors Bradley and Davis are to be commended for their courtesy in employing, in their responses to my chapter, the terminology that I actually developed there. One exception in Bradley's response, however, deserves brief comment because it is indicative of a more general problem. In the context of reflecting on attitudes toward what I have called the robust formational economy concept, Bradley asks, "Why do many scientists like Van Till believe in a naturalistic origin of life?" Well, do I in fact "believe in a *naturalistic* origin of life"? That depends, of course, on the operative meaning of the key term, "naturalistic."

Many terms commonly employed in discussions on natural science and Christian theology have multiple meanings that must be carefully distinguished in order to avoid misunderstanding or misrepresentation. The words "naturalistic" and "naturalism" are among them. In some contexts the word "naturalism" functions as the name for a broad and comprehensive worldview that denies the reality of God. This broad meaning of "naturalism" (or of its associated adjective, "naturalistic") must, however, be distinguished from a narrow, and also commonly employed, meaning that refers only to the idea of focusing one's attention on what can be said about the properties, capabilities, and actions of the natural (I prefer the term, "creaturely") world, without making any commitment about the character or reality of divine action in that same world.

What do I believe concerning the formation of the first life-form in the creation? I believe that only by God's generous provision of the requisite creaturely capabilities for self-organization and transformation could it have come to pass in a way that we would call "naturalistic" in the *narrow* sense of needing no divine interventions to bridge gaps in the creation's formational economy. I shall assume that was the meaning intended by Bradley. The problem is that so many other writers, and their readers, have chosen mischievously to disregard the profound difference between this benign meaning and the pernicious meaning that would follow if the word "naturalistic" were taken in its broad and God-denying sense.

Similar comments could be made with regard to the term "methodological naturalism" as the name of the methodology or strategy ordinarily employed by the natural sciences. If the term "naturalism" were here taken within the limits of its narrow sense, a Christian need not find the term "methodological naturalism" inherently objectionable. However, if in the context of using this term the word "naturalism" had most often been employed in its broad sense to denote a comprehensive antitheistic worldview, then an entirely different message would be conveyed—that the underpinnings of scientific methodology are provided, not by the concept of creation's robust economy of capabilities, but by the presuppositions of a God-denying view of reality. The reader must be relentlessly alert to these subtleties of rhetoric in the science-theology discourse.

Professor Davis, writing from his own perspective of *progressive creationism* (in which episodes of special creation play a key role), clearly wishes to reserve a place for "formational discontinuities"—sudden appearances of new forms by extraordinary means beyond the capabilities of created substances alone—in the formational history of the creation. But I think he does so at too high a cost. According to Davis, these discontinuities are especially important because they "can manifest the *lordship* and *transcendence* of the Creator over the creation." In saying this I think Davis is expressing a very widespread feeling within the evangelical Christian community. The intent is laudable, of course. All Christians should see the Creator's lordship and transcendence manifest in the creation.

But are those Creator-creature relational qualities to be seen only (or even principally) in formational discontinuities? If so, what a tragedy it would be, because the overwhelming majority of phenomena that take place within the creation's formation and continuing operation are characterized by the continuity of creaturely action. Is the Creator's lordship and transcendence to be seen only in the exceptions to this general pattern? Perhaps the proponents of the various versions of special or progressive creationism presented in this book are content with that state of affairs, but I am not. That's one of the reasons for my strong preference for the fully gifted creation perspective. I wish to have reason to celebrate not only God's creativity and generosity, but also his lordship and transcendence in *everything* that the creation is gifted to do, not in some limited number of exceptions that would, if present, call attention to gifts withheld.

Professor Davis asks us to count two particular phenomena as positive evidence of the sort of discontinuities that he has in mind: (1) the appearance of the first living cell, and (2) the "Cambrian explosion" of new phyla during a period of several million years about 600 million years ago. Davis is correct, of course, to point out (as did Bradley) that these are remarkable phenomena for which our present scientific understanding is incomplete (as is also the case for numerous phenomena outside of the arena of biotic evolution). But are they positive evidence for lack of continuity? Not at all. At best they are evidence for lack of understanding.

As I argued earlier, we do not have a positive basis for leaping from evidence for a lack of understanding to the conclusion that there must be gaps in the formational or operational economies of the creation. This is the classic argument from ignorance and no amount of rhetoric will change that. Our ignorance regarding such remarkable phenomena as the first appearance of life or the Cambrian explosion of new forms is certainly not the fault of special or progressive creationists. But these creationists could well be faulted for attempting to advance the argument from ignorance as if it had the probative force to verify the hypothesis of formational discontinuities in the creation.

In closing, I invite the reader to join with me in celebrating the astounding giftedness of the creation as a manifestation of

God's unfathomable creativity and unlimited generosity. And join with me also in experiencing the Creator's lordship and transcendence over the creation, not in exceptions to the creation's giftedness, not in claims for evidence of gifts withheld, not in discontinuities, but in every gift of being that God has given to the remarkable creation of which we are an integral part.

Postscript

FINAL REFLECTIONS ON THE DIALOGUE

Richard H. Bube
Phillip E. Johnson

REFLECTION 1

Richard H. Bube

INTRODUCTION

There are essentially three types of interpretational frameworks that Christians have used to expound the meaning of Genesis.

1. The *completely literal view,* sometimes referred to as the *literalistic view,* contends that God created all things by instantaneous *fiat,* bringing all things from nothing into full being in six twenty-four-hour days about ten thousand years ago. This view has fundamental difficulties both in the area of biblical interpretation and in disagreement with scientific descriptions of earth history. Efforts to support this position have involved such propositions as: (1) the gap theory, suggesting a long period of time between Genesis 1:1 and 1:2; (2) the apparent-age theory, suggesting that the world was created ten thousand years ago but with all the appearances of being much older; and (3) the flood-geology theory, which sought to find explanations of scientific age data through the effects of a worldwide flood. This position rejects the possibility that any kind of scientifically describable natural process (macroevolution) was involved in the origin of different living species. In this book this view is represented by the chapter "Young Earth Creationism."

2. The *essentially literal view* agrees with the completely literal view in holding to the essentially historical character of Genesis 1–3, but allows for figurative nonliteral descriptions to occur in the text. The emphasis is on *harmonizing* the literal biblical text

with scientific descriptions. Two standard variations are: (1) *chronologically accurate age-day theories*, which interpret each day in the Genesis account to correspond to a long period of time during which development occurs by natural process between times of specific *fiat* creation—hence the name *progressive creation*; and (2) *nonchronological day-age theories*, which allow the Creation events of Genesis 1 to be ordered in something other than a chronological framework (topical, liturgical, etc.). This view usually rejects the possibility that human beings came into being as the result of a scientifically describable natural process. In this view current scientific understanding prevents a completely literal view of Genesis, but commitment to a traditional biblical interpretation makes a harmonizing description essential. In this book this view is represented by the chapter "Progressive Creationism."

3. The *essentially nonliteral view* regards the content of Genesis 1 to be a genuine revelation of God, but also considers that any detailed attempt to harmonize this description with a scientific description is misguided, simply because it was not written for the purpose of informing us about twentieth-century scientific theories. In many ways the essentially nonliteral view regards scientific descriptions of origins to be complementary to theological descriptions of origins. In some forms this view may consider Genesis as a kind of divinely inspired story or parable, revealed to make known the fundamental theological truths essential to humankind. The biblical record is seen as revealing the basic truths about creation, but the question of how God accomplished his work of creation is left open, with major insights coming to us by way of scientific developments. Advocates of this position are open to the possibility that all living creatures, including human beings, have historically come into being as the result of scientifically describable natural processes, but are not dogmatic about it. In this book this view is represented by the chapter "The Fully Gifted Creation."

WHAT IS SCIENCE?

What one thinks science can and should do depends to a large extent on what one thinks science is. And, unfortunately, the definition of science is often treated as if it were "up for grabs" by anyone who cares to offer a personal definition. What "scientists or philosophers say" is given the same prominence as what "doing science" actually involves.

One definition that attempts to be true to the practice and purposes of authentic science is this: "Authentic science is *a* way of knowing based upon *testable descriptions* of the world obtained through the *human interpretation* in *natural categories* of publicly observable and reproducible *sense data*, obtained by interaction with the natural world."[1]

The function of this definition is to limit the domain and implications of science appropriately, so that the results of science may be accepted as valid indications of what the natural world *is like*. The definition emphasizes that science does not provide the only way of knowing, gives descriptions of what the world *is like* and not exclusive explanations of what the world *is*, is the result of human activity, is limited by choice to natural categories to define the area of applicability and not because of an atheistic worldview, and involves the testing of human hypotheses by comparison with phenomena in the natural world. To say that something is scientific is not to say that it is absolutely true, but only that its description fits these criteria; to say that something is not scientific is not necessarily to say that it is false, but only that such insight and knowledge comes via other routes than the scientific one.[2] To say that a position is not scientific is a matter of consistent definition and not necessarily a pejorative statement.

Although it is recognized that no human activity, such as is described in the definition of science above, can be a truly objective activity free of all inputs from personal faith systems or philosophical guidelines, it is the goal to carry out science in as objective a way as possible, seeking to understand how the physical world functions "as it presents itself to us," rather than seeking to base our scientific activity on a variety of subjective influences. It is specifically recognized that "authentic science"

---

[1]Richard H. Bube, *Putting It All Together* (Lanham, Md.: University Press of America, 1995).

[2]For a comprehensive collection of papers on various aspects of this subject, see Jitse M. van der Meer, ed., *Facets of Faith and Science*, 4 vols. (Lanham, Md.: University Press of America, 1996). See also the review of this collection by Richard H. Bube in *Perspectives on Science and Christian Faith* 50, no. 1 (March 1998).

cannot be carried out for the purpose of attempting to show the validity of previously assumed philosophical, metaphysical or religious conclusions. Historical examples where scientists have used philosophical perspectives to guide their scientific decisions are usually cases where nonscientific concepts have been used to develop a model, which then can be tested to determine its scientific validity. It is essential to realize that this does not mean that such influences (e.g., moral commitments, theological beliefs, etc.) should not play a general role in (1) guiding what problems to tackle scientifically (e.g., issues of environmental responsibility); (2) deciding in some cases how to tackle them (e.g., testing the effect of pain on human beings cannot be allowed by deliberately inflicting pain); or even (3) suggesting possible theoretical hypotheses that can then be experimentally tested. The essential emphasis is that mechanisms chosen for scientific descriptions must be susceptible to such scientific tests.

Again, this limitation is imposed in order to contribute to the general reliability and acceptability of our scientific descriptions as they are limited by our choice to a partial description of part of reality. If, in the course of such scientific research, we arrive at the point where we cannot find a scientifically acceptable description of an observable phenomenon in the world, we are left with two possibilities: (1) we can conclude that no such description is possible, and that we have encountered a genuine case of God's direct action; or (2) we can conclude that the subject must remain open, since the possibility always exists that further investigation may lead us to a scientific description of God's activity after all.

Whether we are led tentatively to adopt one or the other of these two possibilities, in both cases we are provided with evidence of God's activity when seen through the eyes of Christian faith. It is not as though acceptance of (1) is more consistent with a Christian worldview and commitment than acceptance of (2).

It is, of course, necessary to recognize differences between various branches of what is commonly called "science." Such differences are particularly evident when comparing the physical sciences (e.g., physics, chemistry, geology, and biology) and the social sciences (e.g., psychology, medicine, political science, and sociology). In practice the physical sciences can accommodate the above definition of science more readily, whereas the social sciences, by their very nature, frequently involve major inputs involving human opinions and beliefs as the actual ingredients of the research. Such differences are illustrated by the fact that social sciences have often been separated into two parts: an "objective as possible" approach like that of the physical sciences, and a different approach involving personal relationships and other inputs (e.g., research psychology versus clinical psychology; medical research versus practice of medicine). To ignore this difference between the physical sciences and the social sciences unfortunately contributes further to confusion about the meaning of "science."

A similar issue arises in treating theories in the "natural sciences" (astronomy, cosmology) in which, by the very nature of the situation, it is difficult to perform experiments to test a variety of theories; these theories must often be considered to be essentially metascientific in nature until and unless some kind of testing becomes possible. If indeed no test is possible, there is nothing inconsistent with claiming that God has acted uniquely in a given case to produce a given result, but it is inconsistent to contend that such a claim is part of science.

Over the years, and particularly in the past few years, a major source of confusion in the creation-evolution debate has centered on the definition and purposes of science. It is unfortunate that a false definition of science as the "explanation of everything" has been embraced both by those on the one extreme who hold to an atheistic worldview, and on the other extreme by those who hold to a literal biblical worldview. The person committed to an atheistic worldview argues that since God is not included in scientific explanations, the successes of science show that there is no God. On the other hand, the person committed to a literal biblical worldview argues that the absence of specific divine activity from current scientific explanations is the cause for this lack of evidence for the existence of God, and that therefore the existence of God must be acknowledged by changing the scientific process itself into a "theistic science" so that God's activity can be admitted into it as a scientific explanation.

The problem is centered on the false definition of science as the "explanation of everything," that is, on the assumption that science has the ability to tell us what things *are*. Both of the above groups are misled in adopting this position as a starting point, drawing the conclusion either that science is all we need because of this false definition, or that science as practiced must be changed to justify this false definition for the Christian. *One of the major efforts in the science-theology debate in the future should be directed to removing this false antithesis with its misguided solutions, recognizing from the beginning the deliberate and characteristic limitation of scientific descriptions.*

In order to come to a reliable description of all of the features of reality, both those describable by science and those (such as ethics and personal relationships) requiring disciplines such as theology for their description, the inputs from authentic science and the inputs from authentic theology must be seen as interacting, complementary descriptions, each bringing valid and vital partial insights from their own domains. They cannot be considered as independent or noninteracting, but they must be allowed to guide overall perspectives as the result of an attempt to integrate them in a way that preserves the qualities of authentic description relevant to each.

One who has been involved in the *doing* of science, rather than in formulating or speculating about philosophies generated by scientists, realizes that science does not give us the "explanation of everything," or tell us "what reality *is*," but rather that science gives us "a description of what part of reality *is like*," and therefore at best is capable of giving us only a *partial* description of *part* of reality. There is nothing inherently wrong with the scientific method as historically understood, but there is something wrong with philosophical perspectives that claim the authority of science.

If one wishes to resolve the apparent conflict between science and theology, the way to proceed is by challenging the claim that nonbiblical philosophical conclusions have the authority of authentic science, and not by attempting to reconstruct the scientific method by insisting that God's activity must be inserted into that procedure as a scientific hypothesis or mechanism.

THE IMPORTANCE OF DEFINITIONS

In as complex a subject as the interaction between science and Christian theology—especially in the specific case of creation and evolution—attention to assumed definitions is critical. A major part of the misunderstanding, disagreement, and apparent conflict between different views of creation and evolution can be traced to basic differences in the definitions that are assumed for essential terms and concepts.[3] To introduce this importance of definitions is not to seek to reduce the discussion to gerrymandering or sophistry, as is sometimes charged, but rather to attempt to give authenticity and credibility to the discussion. To illustrate this, let us consider a few of the most critical definitions. The application of these definitions is most consistently set forth in this book in the view described by Van Till.

Limited Terms Versus Worldviews

One of the most fundamental distinctions lies in the recognition of the difference between certain limited words (science, natural, deterministic, chance, creation, evolution) and the general philosophical worldviews extrapolated from them on faith by absolutizing them (scientism, naturalism, determinism, chance, creation, creationism, evolutionism). The addition of "-ism" to many words is a clear signpost that we have moved from the specific to a worldview chosen on faith. We also use capitalization as an indication of a worldview, for example, "Creation" as a biblical worldview and "creation" as a specific process.

Science is a limited human activity as defined above. Scientism is a worldview chosen on faith that matter is all that exists and that science provides the only knowledge of truth possible—it presupposes an atheistic worldview. Christians embrace science and reject scientism.

"Natural" is an adjective describing the types of material and phenomena observable in the physical world and describable, at least in principle, by science—it can be seen as a description of God's regular activity in the physical world. "Naturalism" is a worldview chosen on faith that there is no reality beyond the material physical world investigated by science—it presupposes an atheistic worldview. Christians embrace the natural as God's activity and reject naturalism. To claim that natural materials and processes are responsible for

---

[3]Richard H. Bube, "Penetrating the Word Maze," *Perspectives on Science and Christian Faith* 40 (1988): 104, 170, 236; 41 (1989): 37, 109, 160, 236; 42 (1990): 45, 119, 185, 254; M. Poole, *God and the Big Bang* (N.p.: England: CPO Worthing, 1996).

some observed phenomena is in no sense to claim that God is not involved in the creation, preservation, and functioning of those phenomena. Without the continuing, moment-by-moment, foundational activity of God, such natural phenomena simply would not exist.

No more damaging confusion is caused than that which arises from assuming that calling something "natural" means that God is not involved, and that the involvement of God can be assured only by the treatment of specific phenomena as exclusively "supernatural."

A "deterministic scientific description" is one in which future properties of a system can be accurately predicted from a knowledge of present properties. A worldview of "determinism" asserts that all events taking place in the world are determined to occur, as in the concept of "fate," and that such concepts as "individual choice" or "personal responsibility" are at best illusions or epiphenomena. A "chance" scientific description is one in which only probabilities concerning future properties of a system can be accurately predicted from a knowledge of present properties. A worldview of "chance" asserts that all events taking place in the world happen spontaneously so that such concepts as meaningfulness or God's providence are irrelevant. Since evolution is often described as a "chance" process, this represents an objection for those Christians who hold that a scientific chance description automatically supports a meaningless chance worldview. There is, however, the curious paradoxical situation that neither scientific deterministic descriptions nor scientific chance descriptions are by themselves adequate to describe genuine personal choice and responsibility. Christians recognize the limited utility of deterministic and chance scientific descriptions, and are able to see such descriptions as applicable to God's activity in the world, while rejecting the absolute worldview positions of determinism and chance.

What is essential for our present discussion is the realization that neither "determinism" nor "chance" as exclusive worldviews are ultimately supported by science.

"To create" is a verb implying the bringing into being of something new that did not exist in that form and with those properties before—in a general sense, it is a description of the origin of novelty in the world through the continuing creative activity of God. In principle it could occur either as a continuous process susceptible to scientific description, or as an instantaneous act of God not describable scientifically. "Creation" refers to a foundational biblical worldview based on faith in God, the Maker of heaven and earth, as described more fully in the summary section at the end of this chapter. "Creationism"[4] is a worldview in which the specific mechanism of this activity must be identified with instantaneous, nonscientifically describable acts of God, in a literal and historical way with certain biblical passages; it is assumed that the purpose of these biblical passages is to give us literal information about physical mechanisms. Christians generally embrace the biblical worldview of creation and creation by God through a variety of possible mechanisms; Christians committed to a literal biblical interpretation embrace creationism.

"Evolution" is a description of a possible mechanism that we can use to describe the process of coming into being of something new, which did not exist in that form before, through certain scientifically describable processes—it is in the form of a theory whose suitability for the description of the actual events occurring in the physical world is impressive but incomplete, open to question, and constantly subjected to test. "Evolutionism" is a worldview in which it is assumed on faith that all that was, is, or ever will be is the product of meaningless chance events that result in transforming changes—it is in the form of a faith commitment that assesses all observations and data in terms of an atheistic perspective. Whereas many Christians accept the possibility and utility of evolution as a description of God's activity in creation, they reject evolutionism.

Methodological Naturalism

"Methodological naturalism" has become a battle cry in many science-versus-faith debates.[5] Basically it is because this use of words implies that the methodology adopted in science by choice (i.e., the limitation of scientific descriptions to natural categories) results from or leads to the acceptance of "naturalism,"

---

[4]Ronald L. Numbers, *The Creationists* (New York: Knopf, 1992).

[5]See van der Meer, *Facets of Faith and Science*; Richard H. Bube, *Essay Review* in *Perspectives on Science and Christian Faith;* Alvin Plantinga, *Perspectives on Science and Christian Faith* 49 (1997): 143; R. C. O'Connor, *Perspectives on Science and Christian Faith* 49 (1997): 15

an atheistic worldview and the deliberate rejection of any activity of God in the natural world. *The adoption of a methodology for science that is limited to descriptions in natural categories in no way needs to imply an atheistic worldview of naturalism.*

To the practicing Christian scientist, the limitation of scientific descriptions to natural categories is not at all the necessary result of an atheistic worldview but simply a choice to make it possible for science to be a well-defined and reliable, albeit limited, activity. Nor does the adoption of such a methodology imply that no supernatural or metaphysical inputs are allowed in the formulation of possible theories. What the adoption of such a methodology does insist, however, is that for a particular approach to be considered science, it must be subject to experimental test and shown to be an accurate indication, within limits, of what reality is like, consistent with the definition of science given above.

Consider the different possible interpretations that could be given to the statement that "science today is based on methodological naturalism."

1. The atheist regards "methodological naturalism" as a truism. Since his worldview (chosen on faith) asserts naturalism, it only makes sense to carry out a scientific investigation involving natural categories only. "Science" and "scientism" are identical.
2. The Christian engaged in Christian apologetics sometimes reads the position of the atheist backwards and arrives at the following picture. The application of naturalism to the practice of science requires the adoption of a methodological naturalism that will guarantee that any activity of God is by definition ruled out, not only from the chosen techniques of science but from the fundamental operation of science itself. Therefore, the choice of the practicing scientist to limit himself to descriptions in natural categories is seen as a deliberate choice guaranteed to prevent any supernatural inputs, and hence as an endorsement of naturalism. The only way to challenge the dominance of naturalism over science is to forsake the methodology limiting science to description in terms of natural categories, and to open up a new science in which supernatural categories are accepted as scientific along with natural categories.

3. The Christian involved in doing science may agree that science is based on methodological naturalism, meaning that science is a discipline involving the human interpretation of phenomena in natural categories only. This is a choice made to define the capabilities and limitations of science.

In this context methodological naturalism has no philosophical or theological significance, since the Christian scientist believes in general that *all* authentic scientific descriptions correspond to descriptions of what God's activity in the world appears to be. It specifically does not involve embracing the worldview of naturalism.

The possibility remains open that some phenomena in the world may not be describable in natural categories, that is, that nonscientifically describable phenomena may occur. But the Christian scientist also seeks to maintain a situation where the term "scientific" carries some specific meaning and assurance of validity, and hence insists that nonscientifically describable phenomena should not be called "science." Arriving at this conclusion often leads to open disagreement both with atheistic and Christian philosophers, who often assume that the term "science" appropriately describes a large realm of human thought and experience beyond that described in our initial definition.

Intelligent Design

A major theme in the history of Christian apologetics has been the "argument from purpose and design," also known as the "teleological argument." When one looks at the unique properties of matter and the earth that allow the existence of human life, the marvels of animal and human physiology, and the many examples in the plant and animal world where the existence of one species is totally dependent upon the interactive existence of another, one is struck by the amount of evidence that can be interpreted as indicating that all of this has been designed by a Great Designer. Many of these arguments in recent years have been stated under the title of "the Anthropic Principle"[6] that

---

[6]J. D. Barrow and F. J. Tipler, *The Anthropic Cosmological Principle* (Oxford: Clarendon Press, 1986).

summarizes many of the nuclear, atomic, and gravitational phenomena that appear to be "fine-tuned" to allow the development and/or existence of intelligent life based on carbon.

All such evidences from phenomena in the natural world for the existence of a Great Designer are powerfully consistent for the person who has a personal relationship with God the Creator and Sustainer. The Christian scientist repeatedly marvels at the evidences he sees for the results of God's design in the properties and development of the universe.

Whether they are accepted as evidence for the activity of the God of the Bible, however, depends primarily on the faith commitment of the person who is considering them. It is logically possible to simply dismiss them as a grand "shake of the dice" that happened in a one-in-a-trillion toss to lead to a universe that will sustain human life; if it had not happened, we would not be here to think about it. Or they may be considered as mysterious evidence for the existence of some kind of "life force" with little or no relationship to the God of the Bible.

Related to this is the more recent emphasis upon the concept of intelligent design to be regarded as an appropriate "scientific mechanism" in our efforts to describe and understand the workings of the universe.[7] For the reasons described above, some Christian apologists have concluded that a science based on methodological naturalism is inextricably linked to a worldview of naturalism, and that therefore we must rescue ourselves from this situation by introducing the concept of intelligent design as a mechanism in scientific descriptions.

There is no objection to using the concept of intelligent design as a guide in helping to suggest how to construct suitable models of physical reality, provided that these models are capable of being subjected to test and description in natural categories before they are accepted as scientific. *Intelligent design for the Christian is a general concept underlying all descriptions, scientific and nonscientific, affirming the creative and sustaining activity of God.*

But if the concept of intelligent design is advanced as a substitute for natural categories of description, limiting the specific instances being considered to acts of God's "intervention" in the

---

[7]Michael Behe, *Darwin's Black Box* (New York: Free Press, 1996); William A. Dembski, "Intelligent Design as a Theory of Information," *Perspectives on Science and Christian Faith* 49, no. 3 (1997): 180.

"gaps" in our understanding, and considering intelligent design itself as a valid scientific description, critical harm is done to our concepts of the relationship between scientific descriptions and God's continuing activity in creating and sustaining.

There is the frequent temptation to consider that we can meaningfully decide what God has done and does do, directly on the basis of our understanding of who God is and what God could do. The history of science and Christianity supplies many examples, both in the construction of models of the physical world and in biblical interpretation, where the decision about what God has done has been made incorrectly on the basis of what our presumed knowledge of God would lead us to believe that he has the ability to do. If, in the case of evolution, for example, we wish to answer the question, "How did God achieve his designs in biological development?" we must turn to investigate what it is that God has indeed done, and what form his activity has taken in the actual working out of his creative will. Otherwise we are subject to such classic errors as arguing that the shape of the planets' orbits must be circular because the circle is God's perfect shape, or to arguing that sin must be an illusion because God has made us and God is good.

Second, there is the whole area of interaction between such concepts as "natural law" and "God's intervention" in the world. Many writers speak of "natural law" as though "laws" were self-existing elements that God called into existence to rule the physical world. *Within the area of science, "natural laws" are human descriptions of God's regular creative and sustaining activity. Laws do not cause anything to happen; they are descriptive, not prescriptive.*

For God to act in a way different from this regular creative and sustaining activity—as, for example, in the doing of a "miracle"—he does not have to "break his laws," "set aside his laws," or "intervene in his laws" to accomplish his purpose. Just as we can understand the ordinary "laws" of nature as our descriptions of God's regular activity, so we can understand a "miracle" as our description of God's special activity.

Third, there is the concept of "soul" and its implications for reflections on creation and evolution.[8] There is a growing awareness of the difference between the basic biblical concept of soul

[8]M. A. Jeeves, *Human Nature at the Millennium* (Grand Rapids: Baker, 1997).

as "living self," a concept increasingly supported by growing knowledge of the human being, describing a set of properties of the whole human being, and the "immortal soul," a concept of classical dualistic models of human nature, which has often been used as an argument against the "natural" theory of evolution. It appears to be much more appropriate, both scientifically and theologically, to think of the soul as describing what a person "is," rather than what a person "has."

SUMMARY[9]

It would be unthinkable for any Christian not to confess, "I believe in creation."

The worldview summarized in the biblical doctrine of creation is one of the most fundamental sets of doctrines revealed to us by God. It reveals to us that the God who loves us is also the God who created us and all things, and establishes the identity between the God of religious faith and the God of physical reality. Our belief in creation underlies our trust in the reality of a physical and moral structure to the universe, which we can explore as scientists and experience as persons. Our belief in creation enables us to see that the universe and everything in it depends moment by moment upon the sustaining power and activity of God. Our belief in creation provides the foundation for our faith that we are not the end-products of meaningless processes in an impersonal universe, but men and women made in the image of a personal God. Our belief in "creation out of nothing" affirms that God created the universe freely and separately, and rejects the alternatives of dualism and pantheism. To worship God as Creator is to emphasize both his transcendence over the natural order and his imminence in the natural order. It is to recognize that his mode of existence as Creator is completely other than our mode of existence as created. To appreciate God as Creator is to recognize that the creation is intrinsically good, and that sin and evil do not arise ultimately from properties of that creation such as finitude and temporality. The rationale for scientific investigation, the assurance of ultimate

---

[9]This summary is based on a previously published succinct statement of the various issues involved in the creation-evolution discussion. See Richard H. Bube, "We Believe in Creation," *Journal of the American Scientific Affiliation* 23 (1971): 121.

personal meaning in life, and the nature of evil as an aberration on a good creation are all intrinsic to such an appreciation.

The biblical doctrine of creation plays such a foundational role in all of the biblical revelation that it is unfortunate when the word "creation" is used narrowly and restrictively to refer—not to the fact of creation—but to a possible mechanism in the creative activity. Once again we have the pattern mentioned above, where creation refers to the mechanisms active in the process of creation, and creation refers to the biblical worldview based on the fundamental significance of God's creative activity.

When it is implied that creation and evolution are necessarily mutually exclusive, or when the word "creation" is used as if it were primarily a scientific mechanism for origins, a profound confusion of categories is involved.

The implication is given, deliberately or not, that if evolution should be the proper mechanism for the growth and development of living forms, then creation would have to be rejected. To pose such a choice is to play directly into the hands of those secular philosophers who argue that their understanding of evolution does away with the theological significance of creation. If such a philosopher is wrong to believe that a biological description does away with the need for a theological description, the Christian antievolutionist is wrong to believe that his theological description must make any biological description impossible.

The key to much of the evolution controversy lies in the recognition of the necessity and propriety of descriptions of the same phenomena on different levels of reality.

Even a complete biological description does not do away with the need for a theological description, any more than a complete theological description does away with the possibility of a biological description. Biological evolution *can* be considered without denying creation; creation *can* be accepted without excluding biological evolution. Biological evolution is a scientific question on the biological level; it would be unfortunate indeed if a mistakenly conceived scientific question were permitted to become the crucial point for Christian faith.

Of course, it is important to realize that evolutionary philosophy—shall we rather say "evolutionary religion"—may well be something quite different, leading to a worldview of evolutionism. In its anti-Christian form, such philosophical evolutionism may involve an exaltation of man, a denial of the reality of moral guilt in any theological sense, and hence an interpretation of the life and death of Jesus as nothing more than a good example. In this view, continued development and improvement are inevitably assured as man, now become conscious of evolution, completes for himself the process of the ages. Such evolutionism is a faith-system that competes for the religious allegiance of men, and against which the Christian faith is called to stand. But, if it is true that the evolutionist must realize that he has little scientific support for extrapolating biological evolution into a general principle of life, the Christian antievolutionist must realize that he has little religious justification upon which to attack a scientific theory dealing with biological mechanisms.

How tragic it often is when Christians, seeking to avoid the errors of philosophical evolutionism, promulgate the falsehood that the efficacy of faith in the atonement of Christ effectively depends upon the dogmatic acceptance of creationism and the dogmatic rejection of any evolutionary processes as descriptions of God's activity in establishing creation.[10]

---

[10]I appreciate very much the contributions of a number of friends and colleagues in helping to refine this essay and clarify its intentions.

REFLECTION 2

Phillip E. Johnson

Richard Lewontin, professor of genetics at Harvard University, is one of the most influential evolutionary biologists in the world. Lewontin expressed his views on the relation of evolutionary theory to atheism in a remarkable essay in the *New York Review of Books* in January 1997. I'll start with two of his main points, which illuminate for us what is at stake in the creation-evolution controversy. Then I'll proceed with three more points of my own to show the kind of thinking we need to make progress toward some solutions.

LEWONTIN ON PHILOSOPHICAL MATERIALISM AS THE BASIS OF SCIENTIFIC KNOWLEDGE

Lewontin is skeptical of much of what passes as evolutionary science. In particular, he is downright scornful of the stories that Richard Dawkins and others tell about how natural selection supposedly created complex biological structures like the eye, the wing, and the brain. Like many other scientists, Lewontin calls these accounts "just-so stories" (i.e., imaginative fables like those written by Rudyard Kipling for children), and dismisses the bulk of Dawkins' work as based upon "unsubstantiated assertions or counter-factual claims." So far Lewontin might sound like a creationist, but in fact he considers himself a Darwinist and agrees with Dawkins that some combination of chance events (mutation) and natural law (natural selection) must have produced the wonders of the living world. Why?

The reason is that Lewontin identifies science with a philosophical doctrine called "materialism" (matter is all there is), and considers scientific materialism to be virtually the same thing as rationality. He explains,

> We take the side of science in spite of the patent absurdity of some of its constructs, ... [and] in spite of the tolerance of the scientific community for unsubstantiated just-so stories, because we have an a priori commitment, a commitment to materialism.... Moreover, that materialism is absolute, for we cannot allow a Divine Foot in the door.

In other words, the materialism comes first and the scientific investigation comes second. The priority of materialism implies that the most important task of science educators is not to teach facts, or experimental techniques, but rather to convince their students that only science can tell us how the world works, and materialistic explanations are the only ones that are acceptable in science. In Lewontin's own words,

> The primary problem is not to provide the public with the knowledge of how far it is to the nearest star and what genes are made of.... Rather, the problem is to get them to reject irrational and supernatural explanations of the world, the demons that exist only in their imaginations, and to accept a social and intellectual apparatus, Science, as the only begetter of truth.[1]

That is a concise statement of a philosophical position called "scientific materialism" (or "scientific naturalism"),[2] and I am confident that every one of our authors, including Howard Van Till, would agree that it is equivalent to atheism and hence unacceptable. If science is the only begetter of truth, and science assumes that nature (matter) is all there is, then God is entirely out of the picture. So far everybody agrees.

Now I come to the main point of difference. Creationists (both of the young and old earth sorts) argue that evolutionary

---

[1]Richard Lewontin, "Billions and Billions of Demons," *New York Review of Books*, 9 January 1997, 28–32. All quotations from Lewontin are from this article.

[2]For present purposes, "naturalism" and "materialism" mean the same thing. Both affirm that nature is a closed system of material causes and effects, which can never be influenced by anything supernatural (such as God).

theory is based primarily on naturalistic philosophy, and conclude that Christians should therefore regard its conclusions with suspicion. We respect the expertise of biologists when they tell us what they have observed in the study of biology, but we do not permit them to dictate to us on matters of religion or philosophy. Creationists would say that the most important "facts" that evolutionary science presses us to accept are not facts at all, but inferences drawn from philosophical assumptions that exclude God from reality. In particular, the Darwinian mechanism (mutation and natural selection) has no creative power of the type required to make complex organs, or to supply the immense quantity of genetic information required for a biological cell. If biologists believe in the creative power of natural selection, it is because of their philosophy and not their observations.

Theistic evolutionists part company with us here. They agree that many influential Darwinists identify their science with a naturalistic worldview, but they argue that this identification is spurious. The most important findings of evolution, theistic evolutionists say, are validated by rigorous scientific testing that is independent of any religious bias. We can have confidence that this is so, they argue, because the findings are endorsed by many scientists who are Christians. Theistic evolutionists admit that evolutionary theory is frequently used to promote atheism, but they say this is the fault of individual atheists and does not discredit other evolutionary scientists or their findings. Christians should therefore accept the scientific theory of evolution, including the creative power of the Darwinian method, and contest only the misuse of that theory by atheists like Richard Dawkins.

LEWONTIN ON "THE TRUTH THAT MAKES US FREE"

Lewontin's second major point starts with one of the most famous statements of Jesus. This may seem odd for a self-proclaimed atheist, but Jesus has a way of identifying the issue even for those who reject his teaching. Lewontin begins by noting what he calls a "deep problem in democratic self-governance." What if the voters reject the teachings of science, and trust instead in some other source of knowledge, such as supernatural revelation? Democracy may lead a nation to disaster if the people are irrational and believe false prophets who only want to lure them into slavery. Lewontin explains why education alone cannot guarantee that the people will follow true rather than false teachings:

> Conscientious and wholly admirable popularizers of science like Carl Sagan use both rhetoric and expertise to form the mind of masses because they believe, like the Evangelist John, that the truth shall make you free. But they are wrong. It is not the truth that makes you free. It is your possession of the power to discover the truth. Our dilemma is that we do not know how to provide that power.[3]

The "truth" in this sense is not a set of facts or specific answers, but a starting point for reasoning that empowers us to find the answers as particular problems arise. (Of course, this is also what Jesus meant, but the starting point is himself.) As applied to the issues debated in this book, this means that what we most need to know is not the answers to specific questions (e.g., How old is the earth?), but rather how to think about these matters so that we have a chance to find the truth and to recognize it as truth when we see it.

The concept that the important truths are the ones that empower us to find new truths helps to explain why the leading evolutionary scientists do not think much of the idea that God is our Creator. The reason is not that they think that biological evolution proves that God does not exist. The reason is rather that evolutionary science starts by assuming naturalism, and hence excluding any role for God, and then it builds on that foundation. Whether naturalism is ultimately true or not, it is the starting point that empowered the scientists to discover the truth, if—and that is a very *big* "if"—the orthodox theory of evolution actually is the truth. Perhaps it is possible to interpret evolution by natural selection as God's chosen way of creating, if you have a sufficiently powerful personal motivation for wanting to do that, but absent such a motive, why not stick with the philosophy that enabled you to discover the theory?

From the perspective of scientific naturalism, theists are perennially in retreat. New truths are discovered by employing

---

[3]Lewontin, "Billions and Billions of Demons."

naturalistic assumptions, and then the theists respond by modifying their system to make it unfalsifiable. When scientists who are theists do make a contribution to evolutionary theory, it is because they were able to put aside their theism and think like naturalists. If all theists can do is to defend their position, and if starting from theistic assumptions never leads to the discovery of new knowledge, then theism doesn't have much intellectual value. That is why theism is labeled "religious belief" and confined to the margins of the academic world, and why professors so often are patronizing or even scornful towards Christian students who try to bring Bible-based thinking into the classroom. Sometimes the professors who do this are themselves Christians, but for academic purposes they have adopted the prevailing view that rational thought must be based on naturalistic assumptions.

METHODOLOGICAL NATURALISM

Theistic evolutionists accept naturalism as a methodology in science, but not as a worldview or absolute truth. (Howard Van Till's "robust formational economy principle" is another term for methodological naturalism.) Methodological naturalists do not necessarily say that God does not exist, nor do they necessarily say (as Lewontin does) that science is the only begetter of truth. What they do say is that God may not be invoked as a factor in a scientific explanation. Science confines itself to naturalistic explanations, and the reliability of this research strategy is supposedly confirmed by the demonstrated success of science in providing the technology (airplanes, medicines, the Internet) upon which even religious fundamentalists rely. Religion proceeds by other methods, and any rational form of religion accepts the facts and theories discovered by science. According to theistic evolutionists, science and religion do not conflict because they deal with different subjects and different ways of knowing. Conflict appears only when either science or religion trespasses on the territory of the other, as when Richard Dawkins claims that Darwin made it possible to be an intellectually fulfilled atheist, or when Phillip E. Johnson claims that natural selection has no creative power. The doctrine that methodological naturalism in science is fully compatible with theism in religion is frequently stated in the form of a slogan, such as "The Bible is not a science textbook" or "Religion tells us how to go to heaven and science tells us how the heavens go."

The weak point in the logic of theistic methodological naturalism is that the distinction between "naturalism as a methodology" and "naturalism as a worldview" collapses when science insists on explaining the entire history of the cosmos and, to that end, conclusively presumes that naturalistic solutions are available for every problem. A determination to look for naturalistic processes may be a mere methodology, but an a priori certainty that they always exist has to rest on strong assumptions about reality. For example, evolutionary scientists do not merely say that they are trying to find out whether life can evolve spontaneously from nonliving chemicals. They insist emphatically that they know for certain that chemical evolution can make life, even though the precise evolutionary pathway may as yet be unknown and the experimental results may seem unpromising. If someone suggests that God may have been directly involved in the creation of life, methodological naturalists scornfully dismiss the idea as an attempt to insert a "God of the gaps," who will inevitably be discredited when science discovers a true theory of chemical evolution. This reasoning takes the question of God's possible involvement outside of scientific investigation altogether and makes naturalism unfalsifiable as a matter of faith.

I summed up the logical implications of the strong version of methodological naturalism in my book *Reason in the Balance:*

> If employing methodological naturalism is the only way to reach true conclusions about the history of the universe, and if the attempt to provide a naturalistic history of the universe has continually gone from success to success, and if even theists concede that trying to do science on theistic premises always leads nowhere or into error (the embarrassing "god of the gaps"), then the likely explanation for this state of affairs is that naturalism is true and theism is false.[4]

Christian intellectuals have unwisely taken comfort in the fact that methodological naturalism does not prove that God

---

[4]Phillip E. Johnson, *Reason in the Balance* (Downers Grove, Ill.: InterVarsity Press, 1996), 211.

does not exist, so there is some wriggle room to allow God back into the picture as the undetectable ruler of the natural realm. What the method does do is imply that Christian theism is intellectually uninteresting and unsupported by evidence. That implication has been quite sufficient to make naturalism the ruling philosophy in the universities.

INTELLIGENT DESIGN AND WHY THEISTIC EVOLUTIONISTS OPPOSE IT

Much current discussion of intelligent design focuses upon molecular biologist Michael Behe's book, *Darwin's Black Box*.[5] Behe shows that the invisible world of molecular systems is replete with examples of irreducible complexity, meaning systems composed of many complex parts, all of which have to be present at once for any part to perform a useful function. Such systems cannot be built up part-by-part through the mindless Darwinian process, which (assuming it capable of producing even a part in the first place) cannot preserve a presently useless part in the hope that it will become useful at some time in the future. Although Behe insists that organisms are designed, he does not insist that they were created suddenly. Designers can work step-by-step, and they can gradually change one kind of thing into another. The crucial thing about a designer is that it can look ahead, and so it can put parts in place one-by-one even though the parts are not presently useful. Hence I can assemble a bicycle from a kit, even though the frame and pedals are useless until they are connected to the wheels. To recognize intelligent design in a complex organ or machine is in no sense to depart from science. For example, computers are intelligently designed, but they operate by lawlike processes that are eminently subject to scientific study. Being a computer scientist in no way entails believing that unintelligent material processes can build a computer without a designer.

Behe says at one point that he is not a creationist, at least if that term means someone who is concerned about supporting the creation account in the Bible. He also does not challenge evolution, if that term means "common ancestry." Then why isn't

[5]Michael Behe, *Darwin's Black Box: The Biochemical Challenge to Evolution* (New York: Free Press, 1996).

Behe classified as a theistic evolutionist? He would be if that term meant a theorist who does not rely on the Bible or other religious authority, and accepts gradual development of organisms over long periods of time, but who sees the need for some guiding (i.e., designing) intelligence. The defining characteristic of theistic evolution, however, is that it accepts methodological naturalism and confines the theistic element to the subjective area of "religious belief." It is (barely) acceptable in science to say, "As a Christian, I believe by faith that God is responsible for evolution." It is emphatically not acceptable to say, "As a scientist, I see evidence that organisms were designed by a preexisting intelligence, and therefore other objective observers should also infer the existence of a designer." The former statement is within the bounds of methodological naturalism, and most scientific naturalists will interpret it to mean nothing more than "It gives me comfort to believe in God, and so I will." The latter statement brings the designer into the territory of objective reality, and that is what methodological naturalism forbids.

I believe that Behe's thesis is correct, but I want to emphasize a different point here. For Christian theists, the hypothesis of intelligent design in biology ought to be extremely interesting, even if they prudently withhold judgment while considering all the possible objections. We live in a culture in which most institutions of higher education have gradually shed their Christian roots and embraced naturalism. Christianity remains strong as a popular movement, but its intellectual influence has steadily dwindled since the triumph of Darwinism in the late nineteenth century. No one doubts that the acceptance of evolutionary naturalism in science has been a major force in driving Christianity to the margins of intellectual life. Even if Howard Van Till doesn't think that Darwin made it possible to be an intellectually fulfilled atheist, he knows very well that lots of other people think so.

What if Behe is right? In that case the confident materialists have been misled, and they have built a very proud tower of theory on a foundation of sand. Indeed, a major reason that scientific naturalists are so reluctant to believe that organisms are designed is that the potential consequences are so far-reaching. A great deal of scientific knowledge has been built around the assumption that material processes can do all the work of biological creation. If evolutionary science is wrong on so fundamental a point, it may be wrong on many others. We may be in for a worldview revolution as spectacular as that caused by Galileo—or Darwin. Persons who have boldly promoted scientific naturalism as a worldview would look as foolish as all those Marxists who assured us that the worldwide triumph of their system was inevitable. Christian intellectuals might reasonably worry that such a possibility sounds too good to be true, but wouldn't you expect them at least to think that it was good?

Interest should be all the greater because Behe's scientific credentials are impeccable, and because he is not the only scientist who is describing a widening rip in the supposedly seamless fabric of evolutionary naturalism. Professor James Shapiro of the University of Chicago, who is about equally critical of creationists and Darwinists, paints a scientific picture virtually identical to Behe's. Just to give the flavor of Shapiro's article, here is a string of excerpts from a 1997 article in the *Boston Review:*

> The molecular revolution has revealed an unanticipated realm of complexity and interaction more consistent with computer technology than with the mechanical viewpoint which dominated when the neo-Darwinian Modern Synthesis was formulated.... It has been a surprise to learn how thoroughly cells protect themselves against the kinds of accidental genetic change that, according to conventional theory, are the sources of evolutionary variability.... The point of this discussion is that our current knowledge of genetic change is fundamentally at variance with postulates held by neo-Darwinists.... Is there any guiding intelligence at work in the origin of species displaying exquisite adaptations that range from lambda prophage repression and the Krebs cycle through the mitotic apparatus and the eye to the immune system, mimicry, and social organization?[6]

I don't blame theistic evolutionists for being initially skeptical of suggestions that biology is replete with evidence of that "guiding intelligence at work." Being careful not to believe something just because we want to believe it is the very essence

[6]James Shapiro, "A Third Way," *Boston Review,* February-March 1997.

of the scientific method. What is surprising is that some theistic evolutionists seem to dislike the idea of design in biology so much that they do not bother to conceal their hope that the whole concept is discredited as soon as possible, preferably without a fair hearing. Why?

One reason is that theistic evolutionists have a lot invested in the claim that science and Christian theism are compatible, and the recognition of evidence for intelligent design raises the prospect of a renewed conflict that they wish to avoid. Another reason is that some theistic evolutionists have so successfully incorporated evolution and methodological naturalism into their theology that criticism of these doctrines seems almost heretical to them. Howard Van Till scarcely mentions scientific evidence, but says that intelligent design in biology is unacceptable because it implies that the Creator had to add something to creation after the ultimate beginning. He writes, "I find it theologically awkward to imagine God choosing at the beginning to withhold certain gifts from the creation, thereby introducing gaps into the creation's formational history." This argument bears an interesting similarity to the position of young earth creationists, who also believe, on scriptural grounds, that God did all the creating at the beginning.[7] Of course, any Christian has to believe that God intervened throughout history in the covenant with the Jews, in the Incarnation, and in the continuing work of the Holy Spirit. But did he intervene to finish the work of creation? Only old earth creationists have to answer yes to that question.

WHAT, THEN, SHOULD WE DO?

I confess I am dissatisfied with all the answers that we have at present. The standard materialistic evolutionary theory called neo-Darwinism is the worst answer, because it is based upon materialist philosophy rather than scientific evidence and leads to the absurd conclusion that even our thoughts are the prod-

[7]Biblical creationism assumes that the created organisms were general types that may have had considerable capacity for variation in response to changing environments. It therefore does not exclude evolution of the kind that scientists actually observe, such as the variations that occur in creatures like finches, fruit flies, and peppered moths. Evolution within the originally created kinds or types is entirely consistent with all creationist positions.

ucts of irrational material processes. In that case, why believe anything at all, including the theory of evolution? Theistic evolution at least recognizes God as Creator, but it gives away far too much in agreeing to adopt naturalistic standards of reasoning. If God is real, and not imaginary, it doesn't make sense to assume that the only way to find out how creation occurred is to assume that God had nothing to do with it. Theistic evolutionists, like atheistic evolutionists, naively accept that natural selection has great creative power even though the evidence in no way supports that conclusion, because they are bemused by the philosophy.

Young earth creationism honors the Scriptures and gives specific content to the biblical doctrine that death and suffering entered the world through human sin. If it turned out to be true, some tough theological problems would become a lot easier. But, as Robert Newman shows us, the young earth scenario seems to face insurmountable scientific problems. Paul Nelson and John Mark Reynolds can respond that the young earth camp includes a few distinguished scientists who are working on those problems. That is true, but nothing I have read so far leads me to be optimistic. I state these personal opinions with some diffidence, largely because I am nowhere near as familiar with the crucial geological evidence and radiometric dating techniques as I am with the main issues of biological evolution. Because of these opinions, most people think of me as an old earth creationist; however, I agree with critics of that position that something is awkward about the idea that God stepped in at various undetermined points in an earthly history of billions of years to do some more creating or to inject new genetic information into the biosphere. Show me a better scientific position than old earth creationism and I'm open to persuasion.

Is it discouraging to have to admit at the end that "I just don't know"? I don't find it discouraging in the least, because I look forward to the exciting work we have to do to get to a position where we can hope to get the answers. The problem is that we want to consider the scientific evidence fairly and without prejudice, but it is hard to do that when so many scientists insist on looking at the evidence only through the distorting lenses of naturalistic philosophy. Until we can separate the philosophy from the science and get an unbiased appraisal of what the evidence does and does not show, it is premature to try to come to any firm conclusions. When we do get an unbiased scientific picture, neo-Darwinism will collapse and we will be in the midst of a scientific revolution so profound that everything will look different.

That's where you come in. What the world needs now is not more people who can argue for one of the existing positions, but people who can advance the ball. Take it from here and run with it!

SELECT BIBLIOGRAPHY

The following is a select bibliography of important books relevant to the creation-evolution dialogue. Under "progressive creationism" and "young earth creationism" we list works that deal largely with defending each version of special creation and place general responses to naturalist evolution consistent with both progressive and special creation under the heading of "intelligent design."

YOUNG EARTH CREATIONISM

Aardsma, Gerald. *A New Approach to the Chronology of Biblical History from Abraham to Samuel.* Loda, Ill.: The Biblical Chronologist, 1997.

Brand, Leonard. *Faith, Reason, and Earth History.* Berrien Springs, Mich.: Andrews University Press, 1997.

Lubenow, Martin. *Bones of Contention.* Grand Rapids: Baker, 1992.

Morris, Henry. *History of Modern Creationism.* Santee, Calif.: Institute for Creation Research, 1993.

The Proceedings of the International Conference on Creationism. Pittsburgh: Creation Science Fellowship, 1986, 1990, 1994, 1998.

Remine, Walter. *The Biotic Message.* Saint Paul: Saint Paul Science, 1993.

Woodmorappe, John. *Noah's Ark.* Santee, Calif.: Institute for Creation Research, 1996.

PROGRESSIVE CREATIONISM

Boice, James Montgomery. *Genesis.* 3 vols. Grand Rapids: Zondervan, 1982, 1985.

Geisler, Norman. *Knowing the Truth About Creation.* Ann Arbor: Servant Books, 1989.

Hayward, Alan. *Creation and Evolution.* Minneapolis: Bethany House, 1995.

Newman, Robert C., and Herman J. Eckelmann, Jr. *Genesis One and the Origin of the Earth.* Downers Grove, Ill.: InterVarsity Press, 1977.

Pun, Pattle P. T. *Evolution.* Grand Rapids: Zondervan, 1982.

Ross, Hugh. *Creation and Time.* Colorado Springs: NavPress, 1994.

______. *The Fingerprint of God.* 2d ed. Orange, Calif.: Promise Publishing Co., 1991.

Weister, John L. *The Genesis Connection.* Hatfield, Pa.: Interdisciplinary Biblical Research Institute, 1992.

Wonderly, Daniel E. *Neglect of Geologic Data.* Hatfield, Pa.: Interdisciplinary Biblical Research Institute, 1987.

THEISTIC EVOLUTION

Barbour, Ian. *Religion in an Age of Science.* San Francisco: Harper & Row, 1990.

Bube, Richard H. *Putting It All Together.* Lanham, Md.: University Press of America, 1995.

Peacocke, Arthur. *Theology for a Scientific Age.* Minneapolis: Fortress, 1993.

Van Till, Howard J., Robert E. Snow, John H. Stek, and Davis A. Young. *Portraits of Creation.* Grand Rapids: Eerdmans, 1990.

Van Till, Howard J., Davis A. Young, and Clarence Menninga. *Science Held Hostage.* Downers Grove, Ill.: InterVarsity Press, 1988.

INTELLIGENT DESIGN

Behe, Michael. *Darwin's Black Box.* New York: Free Press, 1996.

Dembski, William, ed. *Mere Creation.* Downers Grove, Ill.: InterVarsity Press, 1998.

Denton, Michael. *Evolution: A Theory in Crisis.* London: Burnett Books, 1985.

Johnson, Phillip E. *Darwin On Trial.* 2d ed. Downers Grove, Ill.: InterVarsity Press, 1993.

______. *Defeating Darwinism.* Downers Grove, Ill.: InterVarsity Press, 1997.

Moreland, J. P., ed. *The Creation Hypothesis.* Downers Grove, Ill.: InterVarsity Press, 1994.

Overman, Dean. *A Case Against Accident and Self-Organization.* Lanham, Md.: Rowman and Littlefield, 1997.

Thaxton, Charles, and Walter L. Bradley. *The Mystery of Life's Origin.* New York: Philosophical Library, 1984.

GENERAL TREATMENTS OF PHILOSOPHY/HISTORY OF SCIENCE AND THEOLOGY AND SCIENCE

Bauman, Michael, ed. *Man and Creation.* Hillsdale, Mich.: Hillsdale College Press, 1993.

Corey, M. A. *God and the New Cosmology.* Boston: Rowman & Littlefield, 1993.

Hasker, William, ed. *Creation/Evolution and Faith. Christian Scholar's Review* (special issue) 21 (September 1991).

Jaki, Stanley. *The Road of Science and the Ways to God.* Chicago: University of Chicago Press, 1978.

Montgomery, John Warwick. "The Theologian's Craft." In *Suicide of Christian Theology*. Minneapolis: Bethany House, 1970.

Moreland, J. P. *Christianity and the Nature of Science.* Grand Rapids: Baker, 1989.

Pearcey, Nancy, and Charles Thaxton. *The Soul of Science.* Wheaton, Ill.: Crossway, 1994.

Ratzsch, Del. *The Battle of Beginnings.* Downers Grove, Ill.: InterVarsity Press, 1996.

Two important associations and their journals are

Creation Research Society, 10946 Woodside Ave. North, Santee, CA 92071 (journal: *Creation Research Society Quarterly*; subscriptions: Glen W. Wolfrom, Creation Research Society Quarterly, P.O. Box 8263, St. Joseph, MO 64508–8263)

American Scientific Affiliation, P.O. Box 668, Ipswich, MA 01938 (journal: *Perspectives on Science and Christian Faith*; subscriptions: 55 Market St., Ipswitch, MA 01938–0668)

Another important journal is *Science and Christian Belief*, c/o Paternoster Periodicals, P.O. Box 300, Carlisle, Cumbria CA3 0QS, UK.

A first-rate journal that should be read by all who are interested in these issues is *Origins and Design*, Access Research Network, P. O. Box 38069, Colorado Springs, CO 80937–9904. To subscribe contact http://www.mrccos.com/arn/orders/ordinfo.htm, call (888) 259–7102, or write to the address just listed. In our opinion, *Origins and Design* is the single, most important journal related to topics of intelligent design, creation, and evolution currently in print. It should be in the homes and libraries of all Christians who wish to keep up with issues in this area.

THE CONTRIBUTORS

Walter L. Bradley received his B.S. degree in Engineering Science and his Ph.D. in Materials Science, both from the University of Texas in Austin. He taught for eight years at the Colorado School of Mines in Metallurgical Engineering before assuming his current position in 1976 as Professor of Mechanical Engineering at Texas A&M University. He served as Department Head of the Mechanical Engineering Department of 65 faculty from 1989 to 1993. He is the Director of the Polymer Technology Center at Texas A&M University. He has received over $4 million in research grants and contracts resulting in the publication of over 120 referred journal articles, conference proceedings, and book chapters. He has served as a consultant for many Fortune 500 companies including Dupont, Dow Chemical, 3M, Exxon, Shell, Texaco, Boeing, Fina, and Phillips Petroleum. He has been elected a Fellow of the American Society for Materials and selected to be a Texas Engineering Experiment Station Senior Research Fellow. In the area of origins, he has coauthored one book, three book chapters, and two journal articles. He has presented lectures on "Scientific Evidence for an Intelligent Designer" on over sixty major college campuses in the United States and Europe.

Richard H. Bube is Emeritus Professor of Materials Science and Electrical Engineering at Stanford University, where he has been a member of the faculty since 1962, and served as Chair of the Department of Materials Science and Engineering from 1975 to 1986. He received his Ph.D. in Physics from Princeton University in 1950. He was a senior member of the research staff at the RCA David Sarnoff Research Laboratories, Princeton, New Jersey, from 1948 to 1962. Dr. Bube has been personally involved in forty-eight years of research on photoelectronic and photovoltaic semiconductors, and is the author of seven scientific books and over 300 research publications. He is a Fellow of the American Physical Society; the American Association for the Advancement of Science; and the American Scientific Affiliation, which he served as President in 1968; and editor of the journal of the American Scientific Affiliation from 1968 to 1983. He has been a Faculty Sponsor for the InterVarsity Christian Fellowship at Stanford since 1962, and is the author of four books, over 130 papers, and 200 book reviews on the interaction between science and Christian theology. At Stanford he led an undergraduate seminar on "Issues in Science and Christianity" for over twenty-five years, and has been a frequent lecturer on science and Christianity on the campuses of over sixty colleges and universities.

John Jefferson Davis is Professor of Systematic Theology and Christian Ethics at Gordon-Conwell Theological Seminary in Hamilton, Massachusetts, where he has taught since 1975. He is the author or editor of eight books, including *Foundations of Evangelical Theology* and *Evangelical Ethics*. His course, "Frontiers of Science and Faith," was a Templeton Foundation science and religion course award winner. His articles on the relationship of Christian faith and the natural sciences have been published in the journals *Perspectives on Science and Christian Faith* and *Science and Christian Belief*.

Phillip E. Johnson is a graduate of Harvard University and the University of Chicago. He was a law clerk for Chief Justice Earl Warren of the United States Supreme Court and has taught law for thirty years at the University of California at Berkeley. Professor Johnson has lectured widely in the United States and Europe on topics related to naturalism and theism, law and morality, intelligent design, and the creation-evolution dialogue. He is the author of numerous articles, a frequent contributor to such publications as *First Things*, *Books and Culture*, and *Christianity Today*. Among his books are *Darwin on Trial*, *Reason in the Balance*, and *Defeating Darwinism by Opening Minds*.

J. P. Moreland is Professor of Philosophy at Talbot School of Theology, Biola University in La Mirada, California. He has four earned degrees: a B.S. in Chemistry from the University of

Missouri, a Th.M. in Theology from Dallas Seminary, an M.A. in Philosophy from the University of California-Riverside, and a Ph.D. in Philosophy from the University of Southern California. Dr. Moreland has authored, coauthored, or edited twelve books, including *Scaling the Secular City, Does God Exist?* (with Kai Nielsen), *The Creation Hypothesis*, and *Love Your God with All Your Mind*. He has also published over thirty articles in journals, which include *Philosophy and Phenomenological Research, American Philosophical Quarterly, Australasian Journal of Philosophy, Metaphilosophy, The Southern Journal of Philosophy, Perspectives on Science and Christian Faith*, and *Faith and Philosophy*.

Paul Nelson is the Robert Boyle Fellow in Theoretical Biology at the Center for the Renewal of Science and Culture (Seattle), and editor of the journal *Origins and Design*. He received a B.A. in Philosophy from the University of Pittsburgh, and a Ph.D. in Philosophy from the University of Chicago, where his dissertation addressed the foundations of the theory of common descent. His publications include articles in *Biology and Philosophy, Origins Research*, and the volume *Mere Creation* (InterVarsity Press, 1998).

Robert C. Newman is Professor of New Testament at Biblical Theological Seminary, Hatfield, Pennsylvania, and Director of the Interdisciplinary Biblical Research Institute there. His doctorate is in Theoretical Astrophysics from Cornell University, and he has an S.T.M. in Old Testament from Biblical Theological Seminary. He has done additional graduate work in cosmic gas dynamics at the University of Wisconsin, in religious thought at the University of Pennsylvania, in hermeneutics and biblical interpretation at Westminster Theological Seminary, and in biblical geography at the Institute for Holy Land Studies, Jerusalem. He is coauthor of *Genesis One and the Origin of the Earth* (InterVarsity Press, 1977), editor of *The Evidence of Prophecy* (IBRI, 1988), and a contributor to Youngblood, *The Genesis Debate* (Nelson, 1986); Kantzer and Henry, *Evangelical Affirmations* (Zondervan, 1990); Montgomery, *Evidence for Faith* (Probe/Word, 1991); Bauman, Hall, and Newman, *Evangelical Apologetics* (Christian Publications, 1996); Habermas and Geivett, *In Defense of Miracles* (InterVarsity Press, 1997); van Gemeren, *New Interna-*

tional Dictionary of Old Testament Theology and Exegesis (Zondervan, 1997); Dembski, *Mere Creation: Faith, Science and Intelligent Design* (InterVarsity Press, 1998); and Evans and Porter, *Dictionary of New Testament Background* (InterVarsity Press, 1998).

Vern Sheridan Poythress is Professor of New Testament Interpretation at Westminster Theological Seminary in Philadelphia, Pennsylvania. He holds a B.S. from California Institute of Technology, Ph.D. in Mathematics from Harvard University, M.Div. from Westminster Theological Seminary, Th.M. in Apologetics from Westminster Theological Seminary, M.Litt. in New Testament from Cambridge University, and Th.D. in New Testament from the University of Stellenbosch, South Africa. He has written numerous books and articles, including material on the relation of the Bible, theology, and science.

John Mark Reynolds is the founder and director of the Torrey Honors Institute at Biola University. Torrey is a highly competitive great books program that trains students in the classical tradition. Reynolds earned a Ph.D. in Philosophy at the University of Rochester. He has written and spoken widely on the subject of religion and science. He is the editor for the philosophy section of the International Conference on Creationism.

Howard J. Van Till is Emeritus Professor of Physics at Calvin College, located in Grand Rapids, Michigan, where he taught both physics and astronomy for more than thirty years. A graduate of Calvin College, he received his Ph.D. in Physics from Michigan State University. His professional research experience includes both solid-state physics and millimeter-wave astronomy, and he is a member of the American Astronomical Society. His deep desire for encouraging the Christian community to become better informed regarding the character of the creation and of its formational history have led him to write several books and essays in the faith-science arena. He is the author of *The Fourth Day* (Eerdmans, 1986) and editor/coauthor of *Science Held Hostage* (InterVarsity Press, 1988) and *Portraits of Creation* (Eerdmans, 1990). In February 1999 he is being honored with the Faith and Learning Award given by the Calvin Alumni Association.

PERSON INDEX

SUBJECT INDEX

Consider All Points of View with Counterpoints!

Four Views on the Book of Revelation

Is the book of Revelation a prophecy fulfilled when the temple fell in A.D. 70, a metaphor for the battle between good and evil raging until Christ's return, a chronological record of events between Christ's ascension and the new heavens and new earth, or an already/not yet prophetic dichotomy of fulfillment and waiting?

Four Views on the Book of Revelation will expand your understanding of the Bible's most mystifying book, and it may even change your mind.

Moody professor C. Marvin Pate mediates the lively dialogue and adds his own view to those of Kenneth L. Gentry Jr., Sam Hamstra Jr., and Robert L. Thomas.

ISBN: 0-310-21080-1

COUNTERPOINTS COVER KEY DOCTRINAL ISSUES!

Five Views on Law and Gospel

Examine the relevance of the Old Testament law to Christian life with five scholars representing the most common evangelical Christian views.

Douglas J. Moo presents a modified Lutheran answer, Wayne G. Strickland defends the Dispensational view, while Walter C. Kaiser Jr. argues that the "weightier issues" of the Mosaic Law still apply to Christians. Willem VanGemeren and Greg Bahnsen present Theonomic and Nontheonomic Reformed points of view.

ISBN: 0-310-21217-5

Five Views on Sanctification

Five Views on Sanctification will deepen your understanding of the moral perfection of Christ as it presents differing ideas about how that perfection affects the Christian life.

Can we live up to the Wesleyan ideal of "entire sanctification" held by Melvin E. Dieter, or are Reformed scholars like Anthony J. Hoekema correct that complete holiness in this life is impossible? What role do Pentecostals like Stanley M. Horton say transforming gifts of the Holy Spirit play? What is the Keswick tradition of J. Robertson McQuilkin, and what wisdom on this issue has John F. Walvoord found in the writings of St. Augustine?

ISBN: 0-310-21269-3

Available at your local Christian bookstore

COUNTERPOINTS COVER IMPORTANT APOLOGETIC ISSUES!

Four Views on Salvation in a Pluralistic World

If you've ever asked:

What is the eternal fate of indigenous people Christian missionaries have not reached with the Gospel?

How could believing in Christ not be essential for salvation when Jesus said, "No one comes to the Father except through me"?

Then you need to read this Counterpoints volume.

John Hick, Clark Pinnock, Alister E. McGrath, and R. Douglass Geivett and Gary W. Phillips defend their positions on the most challenged belief in Western Christianity.

Dennis L. Okholm and Timothy R. Phillips, General Editors

ISBN: 0-310-21276-6

Four Views on Hell

Learn why not everyone agrees that a hell of fire and sulfur exists or that the unsaved will suffer there forever. Leading scholars discuss whether hell is a literal fire, a metaphor for separation from God, a purgatorial place of purification, or a place of temporary punishment for those who have not accepted Christ.

With John F. Walvoord, William V. Crockett, Zachary Hayes, and Clark H. Pinnock.

William Crockett, General Editor

ISBN: 0-310-21268-5

We want to hear from you. Please send your comments about this book to us in care of the address below. Thank you.

ZONDERVAN™

GRAND RAPIDS, MICHIGAN 49530

WWW.ZONDERVAN.COM

Stanley N. Gundry (S.T.D., Lutheran School of Theology at Chicago) is Vice President and Editor-in-Chief at Zondervan. He graduated summa cum laude from both the Los Angeles Baptist College and Talbot Theological Seminary before receiving his Masters of Sacred Theology from Union College, University of British Columbia. With more than 35 years of teaching, pastoring, and publishing experience, he is the author or coauthor of numerous books and a contributor to numerous periodicals.